BIO — DYNAMIC GARDENING
AND FARMING

*Agriculture is a more natural art
than politics because it co-operates
with nature.*

Plato

*No occupation implants so speedily
and so effective a love of peace
as a country life.*

Plutarch

Ehrenfried E. Pfeiffer
(1899 - 1961)

BIO—DYNAMIC GARDENING AND FARMING

ARTICLES

BY

EHRENFRIED PFEIFFER

MERCURY PRESS
SPRING VALLEY, NEW YORK

This edition of
Bio-Dynamic Farming and Gardening, Vol. III,
first published by MERCURY PRESS in 1984,
is the first publication of Dr. Pfeiffer's
articles in this arrangement.
They were most recently published
in forty-four installments in
the periodical Acres U.S.A.

Permission for this publication
granted by Adelheid Pfeiffer.

Cover design by Peter van Oordt

ISBN   0-936132-67-1

Printed in the United States of America
MERCURY PRESS
Fellowship Community
241 Hungry Hollow Road
Spring Valley, N.Y. 10977

CONTENTS

ACKNOWLEDGEMENTS

We would like to thank Adelheid Pfeiffer for granting Mercury Press permission to publish these papers in book form and we remember in gratitude the late Erica Sabarth who gathered them all with the hope that they would be published together.

We gratefully acknowledge the permission granted by the editors of the journal *Bio-Dynamics* to reprint the following chapters: Chapter I from Vol. VII, No.1, 1948; Chapter II from Vol. X, No.1, 1952; Chapter III from Vol.IX, No.3, 1951; Chapter IV from Vol.XII, No.2, 1954; Chapter VI from Vol. XI, No.1, 1953; Chapter VIII from Vol. XI, No.4, 1953; Chapter X from Vol.II, No.2, 1943; Chapter XII from Vol.IV No.2, 1946; Chapter XV from Vol. III, No.2, 1944. We also gratefully acknowledge permission granted by the editors of *Organic Gardening* for the publication of Chapter IX from the May 1948 issue and Chapter XI from the April 1948 issue. Finally we are grateful for the permission to print Chapter XVI from *The Golden Blade*, 1957, and Chapter XVII from *The Forerunner*, Vol.3, No.1, 1942.

FOREWORD

Despite the increasing number of publications on Bio-Dynamic gardening and farming, it seems appropriate to go back occasionally to the work of one of the great pioneers in this endeavor, Dr. Ehrenfried Pfeiffer. With this volume we present the last of three volumes of articles by Dr. Pfeiffer, the first two having been published in 1983. A study of Dr. Pfeiffer's publications reveals not only the work of a pioneer in the attempts to rejuvenate our dying soils and to improve the nutrition of mankind, but actually the work of an expert whose efforts remain as valid today, twenty-two years after his passing, as they were then. One could say then that he was not only a pioneer, not only an expert, but also a prophet because much of what he foresaw and battled against is here today. Another attribute I would like to add is that of an inspired warrior battling to reclaim the earth for man. It is this last attribute which, to my mind, makes Dr. Pfeiffer's work seem so immediate; it imbues his style with a freshness, a vitality and an urgency which cannot but continue to inspire us to join in the necessary tasks.

I cannot resist here to conclude by quoting from the periodical *Acres U.S.A.* , which in 1983 completed the serial publication of these same articles in order to expose ''a new generation to the works of a giant with the realization that when we progress, we are all standing on the shoulders of giants.'' The series ends with the following editorial comment: ''So ends the *Acres U.S.A.* presentation of the Pfeiffer Papers. These papers deserve a place in the library of every university in the country and on the bookshelf of every farmstead.''

Gerald F. Karnow M.D.
Fellowship Community
Spring Valley, N.Y.

CHAPTER I

A Preliminary Report on Humus Formation in the Root Areas of Plants Under Different Soil Management

There is an active inter-relationship between plants and the soils on which they grow. The crops and their roots do not only take away minerals through absorption, but they dissolve soil, change the soil reaction, add organic matter to the soil, alter the physical structure, and may even influence the entire microflora and fauna of the soil by means of growth hormones. It has been observed that the root areas of certain plants are stimulating to the germination of the seeds of others (oak to pine, spruce and fir; rye to pansy), while the root areas of others have an inhibiting effect. Another observation is that more earthworms can be found in the root areas of clover and of stinging nettle than in those of wheat. There are definite differences in acidity, types of humus as well as of microflora around different species of plants.

The statement of the applied Liebig theory, that what plants take away from the soil has to be replaced, must be amended. That which plants change in the soil or even add to it, good or bad, demands investigation. Humus losses occur with one-sided agricultural methods (chemically and physically) and with one-sided crop rotations. If there were no humus losses, there would be no erosion problems, for humus holds the soil.

In the Biochemical Research Laboratory of the Bio-Dynamic Farming and Gardening Association, we are trying to find out more about the interrelationship of certain crops to soil and humus formation and to humus consumption.

For a change, we did not start with a poor, exhausted soil and then add fertilizer to demonstrate how much better things can grow (ignoring the humus situation). Instead, we began on a soil with a good humus content, which was under sods and had not been cropped for several years—a well rested, natural soil. When we started out this soil had 3.07% organic matter and a pH of 5.8.

We tried to follow up the influence of the plant roots in their respective areas upon the humus content, acidity and availability of minerals. Parallel with the studies of the root areas we also studied fallow areas where no plants were growing.

The test plot was divided in equal halves. The one half was given a mixture of 2/3 manure and 1/3 compost, bio-dynamically treated. The other half was fertilized with a 5-10-5 formula (NPK). The amounts of N, P, and K present in the bio-dynamic manure and compost were calculated so that the dose given was comparable to the 5-10-5 formula.

The vegetables were all sown and the plants all set out on the same day. The biodynamic compost and manure and the chemical fertilizer were also both applied on the same day, just before the sowing and planting. Aside from fertilizer, both halves were treated equally as to cultivation, weeding, etc.

After the plants had grown to the first stage of maturity, soil samples were taken from the root areas of the respective plants; at the same time samples were taken from the fallow areas of both halves of the test plot. All of these samples were tested for pH and organic matter content. The changes in organic matter content and acidity are shown in the following table:

Soil of Test Plot *(before any treatment):*			Organic Matter pH		3.07% 5.8
		Mineral		**Bio-Dynamic**	
Soil of Test Plot: Fallow Area					
(Samples taken at same time as	O.M.	2.07%	O.M.	3.4%	
area samples from plants):	pH	5.5	pH	6.5	
Soils from Root Areas:					
Rendergreen Bush Beans	O.M.	2.27%	O.M.	4.77%	
	pH	6.5	pH	7.0	
Marglobe Tomatoes	O.M.	1.8%	O.M.	4.25%	
	pH	6.8	pH	7.1	

Crosby's Egyptian Beets	O.M.	2.87%	O.M.	5.0%
	pH	6.0	pH	7.2
Romaine Head Lettuce	O.M.	2.72%	O.M.	3.42%
	pH	6.0	pH	6.5
Chateney Carrots	O.M.	2.62%	O.M.	3.47%
	pH	6.0	pH	6.6
Green Peppers	O.M.	1.4%	O.M.	2.67%
	pH	5.8	pH	7.0
Golden Bantam Corn	O.M.	1.77%	O.M.	2.7%
	pH	6.5	pH	6.8
Bibb's Head Lettuce	O.M.	3.67%	O.M.	3.65%
	pH	6.0	pH	6.3

The remarkable thing about these findings is the variation in maintaining the humus content, improving or decreasing it, by the different crops with the biodynamic or mineral fertilizing. The figures almost speak for themselves without need of further comment. However, insight into the problem of crop rotation can also be obtained from a further consideration of them. On the basis of such findings one would just naturally follow a humus consuming crop (corn or peppers) with a humus restoring crop (beans or beets). Nobody doubts that a mineral fertilizer can cover mineral deficiencies as long as the balance of the soil is not disturbed. But in order to grow humus we need organic matter and microlife. And without humus, even the minerals cannot come to full effectiveness. Then too, mineral losses are much higher in soils poor in organic matter.

The difference between the two soil types created in this experiment (the humus-losing and the humus-gaining) was apparent all the way through. It showed up in the root size and formation, in the nitrogen nodules on legume roots, in the health of the plants and even in the carotene and vitamin A content.

Our study of the vitamin A and carotene content is not yet complete. However, with the chromatographic extraction method, we have gotten chlorophyll and carotene columns which vary considerably and in favor of the products grown on the soil which is richer in humus.

3

It may be mentioned here that Dr. G. Howell Harris of the University of British Columbia has recently published data which show variations in sugar and vitamin content of carrots grown in different soils. The sugar content was highest in sandy clays and light clay loams, lowest in heavy clay. The largest roots grew in peat soils although they were the most acid. The vitamin A content was highest in clay soils (when the sugar content was lowest). Sandy loam produced the highest vitamin C content as well as the highest dry weight of the roots. Potassium fertilizer produced the highest sugar content. The balance of all the minerals in the carrots was very little influenced or altered by the fertilizers used, except when a mineral fertilizer with a high nitrogen content was added, then the mineral content of the roots was slightly lowered as was the sugar and vitamin A and C content. When these experiments are placed beside our own observations they really make us stop and think.

This reporter does not necessarily believe that the sugar content alone of carrots should be held as conclusive evidence of their higher food value. The vitamin A content of high sugar carrots may be low. Then we know that people don't like too sweet a carrot, they prefer a fresh aromatic flavor, with good reason. Add to this an observation made in connection with recently published experiments with the intake of radioactive substances (in this case radioactive phosphorus): The proportion of minerals in plants and their intake by plants is not always and not necessarily a direct function of the mineral fertilizer applied.

The sensitive crystallization method, which has been developed by the writer and his coworkers in the course of 20 years, proves valuable in following up the effects of differently treated soils upon plants. The method consists mainly in the crystallization of a 5 or 10% solution of chloride of copper to which small amounts of highly diluted plant tissue extracts are added. The solution is spread out in a thin film an a perfectly clean glass plate, 4" in diameter (provided with a rim), and crystallization takes place in from 14 to 16 hours through evaporation of the solvent at 50-55 relative humidity and a temperature of 80 degrees F. An air-conditioned, temperature and humidity constant cubicle, free from vibrations, is used for the crystallization. The pattern of the crystallization of copper chloride is changed in a characteristic manner by the minute additions. Fine

4

rays of crystals radiating from one center on the plate is a symptom of normal, healthy, undisturbed growth; while deficiencies, abnormalities and disease patterns show up in the formation of many centers, broken off crystals and distorted arrangement. The interpretation of the form pattern is facilitated by comparison with other tests of the same material made with different methods, for instance, chemical analysis, vitamin content, feeding tests, pathology, etc. More detailed information about the method can be found in the literature cited at the end of this article.

In the case of the plants-grown during the test reported here, routine crystallizations were made of each group. The pattern of health shows the single-centered, radiating crystallization, while the pattern of disturbance and deficiency shows the many-centered crystallization with short crystals. All the biodynamically fertilized products show the one-centered pattern; the mineral fertilized products show the distorted pattern.

These illustrations speak for themselves and really need no further explanation. The crystallization method lends itself very well to the study of the degree of health and nutritive value of plants grown under different soil conditions. (See pages 6, 111, 112)

Other Observations on Humus

There is a seasonal variation of humus content, which follows the varying conditions of moisture, warmth and drought. Hot, dry seasons consume more humus, while cool, moist seasons preserve humus. In early summer and late fall, towards the ripening time, when the soil is well shaded by plants or by mulching, we observed an increase of humus; while during the main growing season humus was consumed. It is important that young plants are supported by a good humus content in order to grow "Healthy". As soon as the plants stop growing, but are left in the soil, in humus conserving seasons one will observe an increase of humus. When the plants are harvested and the soil lies unprotected (in stubble, for example) one observes further losses of humus during hot, dry periods. When the evaporation and absorption of moisture are balanced, the soil life builds up a natural increase of humus even without additions of fertilizer or compost, i.e. when all the conditions for the production of living soil are present.

All six photographs made with same concentration 10% Copper Chloride, 4cc + 4cc of a 25% leaf or fruit extract in water.

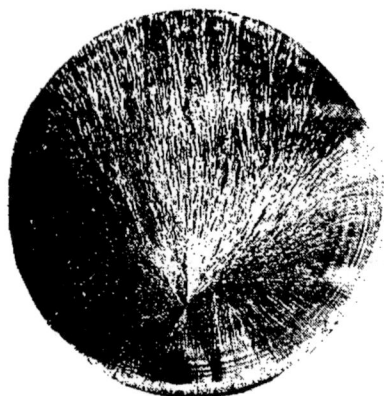

B-D. Bean Leaves

M. Bean Leaves

B-D. Pepper, Fruit

M. Pepper, Fruit

B-D. Tomato, Fruit

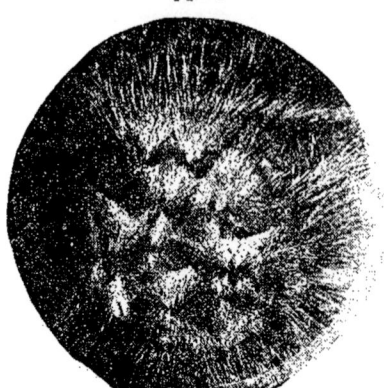

M. Tomato, Fruit

6

This year, with a relatively cool summer and sufficient rainfall (except in September) was particularly humus conserving in the vicinity of the experimental field. The field is well sheltered against drying winds and burning sun by surrounding woods. Had the garden been in an open plain or had the season been one of drought the situation might have been quite different. It was, therefore, due to the location and the weather conditions that the humus content in the M plots caught up at harvest time and thereafter. However the increase was considerably higher in the B.D. plots than in the M plots. The following table gives the figures:

	%of Humus B.- D.		M.		pH(Acidity) B.- D.		M.	
	Smr.	Fall	Smr.	Fall	Smr.	Fall	Smr.	Fall
Beans	4.77	5.87	2.27	3.8	7.0	6.4	6.5	6.3
Pepper	2.67	5.75	1.4	3.57	7.0	6.7	5.8	6.1
Tomatoes	4.25	5.8	1.8	2.6	7.1	6.8	6.8	5.9
Corn	2.7	3.55	1.7	3.6	6.8	6.2	6.5	5.8
Beets	5.0	5.65	2.87	4.05	7.2	6.25	6.0	5.8
Carrots	3.47	4.95	2.62	4.1	6.8	6.8	6.0	5.9
Soil(fallow)	3.4	3.4	2.07	2.87	6.5	6.4	5.5	5.5

As the table shows there is a slight fluctuation in acidity, the organic soil tending more towards the neutral state than the minerally treated soil.

Formative Forces In Crystallization, by E. Pfeiffer, Anthroposophic Press, New York 1936.
Sensitive Crystallization Processes, by E. Pfeiffer, Dresden 1936.

CHAPTER II

Green Manuring, Organic Fertilizer, Compost in Spring?

This question has often been discussed in scientific and agricultural literature and asked by our readers ever and again: Why does green manuring not always give the desired results when the crop is turned under in Spring? Some people even go so far as to say that one should not apply compost or other organic fertilizer, as well as green manure crops in Spring because they cannot act, since they tie down the soil bacteria until they are sufficiently decomposed.

There is a certain amount of truth in the observations which have led to this opinion, but the difficulty can easily be overcome if one knows about the proper decomposition of organic materials in soils and in compost. We have long made a differentiation between raw organic matter and digested organic matter. For many years, the writer has urged that only that fraction of organic matter which has been digested by soil organisms and completely decomposed be called humus.

Green manure crops which are plowed or disked under in Spring take time to decompose. During this period they demand the action of the soil life and bacteria and these are thus not able and available to activate the soil minerals until and unless the green manure is decomposed. The beneficial effect of the green manuring and compost is therefore only felt after some time elapses, frequently only by the late Summer crops or even not until the following growing season. The effect of green manuring is slow to appear but it also lasts longer. Experimental stations have found that its good effects are most evident in the second and third years. This rule also applies to the use of fresh manure and partially decomposed compost which still contain a high fraction of raw organic materials.

The problem is best illustrated if one understands what happens to nitrogen in the soil. Even though green manure crops, especially legumes, contain a large amount of organic nitrogen, the nitrogen bacteria in soil will not work on the soil but on the masses of green material. Hence temporary nitrogen deficiencies appear and the crop immediately following is inadequately nourished. Some twenty years ago, the writer plowed under a crop of winter vetch on heavy, moist, clay soil and planted potatoes thereafter. The results were very poor in spite of the fact that all the necessary plant food was present. The tying down of the nitrogen bacteria during this period of decomposition is sometimes overcome by the addition of nitrogen fertilizers. In fact, when stubble or straw is plowed under, the orthodox school of agriculture recommends adding nitrogen in order to speed up the decomposition. Similar problems exist in compost piles and it has been stated that unless the ratio of carbon to nitrogen is nearly 11 to 1, successful decomposition cannot be obtained.

In our bio-dynamic practice, we acknowledge the truth of these observations, but point to the fact that the decomposition of green manures and compost can be speeded up before the next crop is sown or the compost is applied. One reason why we add the bio-dynamic preparations or the bacterial compost starter (BD Starter) which has been developed recently by the Biochemical Research Laboratory, Threefold Farm, Spring Valley, N.Y., is that we want to *predigest* the green manures and compost before the seeds are sown in the soil. Completely humified compost does not tie down soil life and its nutriments are immediately available. The fact that the quantity of immediately available minerals is many times higher in a treated compost than in an untreated one can be demonstrated by chemical tests. No nitrogen is tied down when the BD Starter is used and the green manure is broken down much more quickly, long before the seedlings begin to grow and require plant food.

One method of applying the BD Starter has recently been practiced and found to be very valuable in California. There the starter is distributed by the Compost Corporation of America in Oakland, and is sprayed over green manure crops and crop residues and stubble, etc., just prior to disking them under. These disked-under crops and crop residues in turn decompose very rapidly under the influence of large numbers of soil and breakdown bacteria. Growers in the salad

9

bowl of the country, in the Salinas area, are very enthusiastic about the results. They have observed that the field which received the starter could be worked much more easily, required three days less labor in preparation for the next crop than the control field worked as usual without the starter application. Lettuce plowed under decomposed within a few weeks and barley stubble took just half the usual breakdown time. This accelerating effect was so noticeable that the starter is now sprayed by airplane where the topography permits its use.

Crop dusting airplane companies have taken up this task and developed special methods so as to bring the starter right to the surface of the field. The pilots fly only one or two feet above the ground. The plane has a spray rig which forces a fine mist of the bacterial spray into the soil and the farmer waits with his disk right on the edge of the particular field in order to dig under the green material, stubble, etc. immediately. In this way nothing is lost and the maximum effect is obtained. One result is that the stubble and straw disappear completely within a few weeks and there are no remainders of it to handicap cultivation and seeding.

The photographs shown here illustrate the contrast between two sections of a lettuce field where barley stubble had been plowed under six weeks previously. On the field section treated with the bacterial starter very little if any stubble is left, the soil has become smooth and friable, while on the untreated control section much stubble remains and has impaired the regular seeding and catch of the lettuce. More than 2,000 acres have been treated successfully in this way. The same results have been observed when plowing under vetch, broccoli and cabbage.

How powerfully the Starter spray acts through its balanced and rich content of its decomposition and soil bacteria can be gathered from the fact that only 2 to 4 oz. of the cultures are used per acre. When sprayed by airplane these are diluted in 5 gallons of water per acre, and the specially rigged plane can cover 40 acres in thirty minutes. The cost is reasonable: $5.50 per acre sprayed inclusive of the Starter.

An entirely new industry has been created in this way, and the idea that green manures or partially decomposed organic materials can tie up soil life and nitrogen belongs to the past. Many years of research

10

and observations have contributed to the development of this new process. The additional starter treatment has also proved its worth in the decomposition of organic wastes, garbage from municipal sources and manures. City garbage is decomposed in the brief time of 2 to 3 weeks, without bad odors, without tying down nitrogen. In fact, the nitrogen fixing bacteria work with vigor and enthusiasm on such composts and even increase their nitrogen content. Predigestion makes possible the most efficient and effective use of composts. Where airplanes cannot be used, in hilly territory and on small farms and gardens, the starter can be sprayed by hand or with a ground rig. In such cases the amount of water for dilution is adjusted to the implement used for spraying, anywhere from 5 to 50 gallons per acre. At some later date further data and information about this problem will appear in these pages.

CHAPTER III

Compost and Fertilizer on Peas
(A Laboratory-Greenhouse Experiment)

In this experiment, the early growth of garden peas was studied in relation to varying amounts of compost and fertilizer applied to flower pots in which the peas were sown. The purpose of the test was to demonstrate the differences in root growth, legume bacteria development, and in the initial growth of the whole plant; also to determine what amount of compost to recommend for optimum results. Since we are using a plant belonging to the legume family, a compost with a low nitrogen content was selected.

Each group of the test consisted of 6 plants grown in 7 inch pots, 3 plants to each pot. The earth was an otherwise untreated soil of good quality with a moderately high organic matter content, not previously used for intensive cropping. The same soil was used for all groups, after being thoroughly mixed to obtain a uniform basis. Compost was mixed with the soil at the rate of 1, 2, 5, 8, 12 and 15 tons per acre. One group was treated with compost as a mulch at the rate of 17 tons per acre, another group received compost at the same rate in the planting holes just before the seeding. For comparison, a 6-8-4 fertilizer was used at the rate of 500, 800 and 1000 lbs. per acre. A control group was grown in the same soil with no treatment whatever. Each pot contained 4 lbs. of soil.

The peas were sown on January 26th and growth was interrupted after 66 days. The plants were carefully removed from the pots, the earth washed away with great care so as not to break the roots. The length of the green plant parts was measured and they were weighed separately from the roots. The length and weight of the roots were also determined. Before weighing, the roots were air dried after all earth particles had been removed. Special consideration was given to the legume bacteria nodules. The length of the roots bearing nodules was measured, the total number of nodules was determined, then the larger nodules (more than 1/2 mm in diameter) were counted and weighed.

At the time their growth was interrupted, most of the plants had developed two blossoms, one additional bud, and had begun to set pods. The pots had been placed in a greenhouse and all received the same amount of light and warmth. Each pot was supplied with a saucer, watering was done from underneath, each receiving the same amount of water. No "drought" condition was created.

The compost used was a rather earthy, well-rotted, bio-dynamically treated mixture, made from garbage and leaves (1/3), manure (1/3) and earth (1/3). It had been kept in a pit during the winter, exposed at all times to the weather. It analyzed as follows: moisture 40%, total nitrogen 0.4%, total phosphate 0.5%, total potash 0.23%, total organic matter 11.1%. In plain language, it was a low NPK compost.

The following table reports some of our findings:

Kind of Treatment	Average Weight in gm. per Plant	Total Number of Nodules on All Roots (6 plants)	Weight of All Nodules over ½ mm diam. in mgm.	Lbs./Acre Applied		
				N	P	K
1 Ton/Acre of Compost	14.4	422	450	8	10	4.6
5 Tons/Acre of Compost	16.3	790	900	40	50	23
15 Tons/Acre of Compost	16.1	696	900	120	150	69
500 Lbs/Acre of 6-8-4 Fertilizer	13.2	577	500	30	50	20
800 Lbs/Acre of 6-8-4 Fertilizer	13.0	640	730	48	80	32
1,000 Lbs/Acre of 6-8-4 Fertilizer	9.4	470	450	60	100	40
Control. Soil Not Treated	11.6	419	600	-	-	-

The basic soil used in the experiment analyzed: pH 6.5, organic matter 2.7%, available potash 300 lbs/acre, available calcium 1250 lbs/acre, available magnesium 10 lbs/acre, available nitrate nitrogen 22 lbs/acre, ammonia trace, available phosphates 150 lbs/acre.

Interpretation and Description of Other Findings:

The maximum growth was obtained with 5 tons of compost per acre.

The maximum number of legume bacteria nodules and the maximum weight of the nodules over 1/2 mm. in diameter were obtained with 5 tons of compost.

The minimum growth was obtained with 1000 lbs. of 6-8-4 fertilizer per acre.

The minimum number of nodules was produced on the control. The minimum weight of nodules was obtained with 1000 lbs./acre of 6-8-4 fertilizer.

Increasing amounts of NPK in the fertilizer application had a marked depressing effect upon the amount and weight of nodule development as well as upon the root growth. Less fine hair and side roots had developed with increasing amounts of NPK fertilizer.

The optimum effect of the 6-8-4 fertilizer was obtained with an application of 800 lbs. per acre.

The densest root development was observed where compost was applied as a mulch and used in the planting holes.

No depressing effect was observed with the application of increasing or even excessive amounts of compost. The optimum effect of 5 tons per acre, however, was not surpassed by applications of 15 or 17 tons per acre of the compost. The practical farmer will be able to understand, on the basis of experience, our observation that no further increase in yield can be expected after the optimum amount of compost has been applied (in the case of this experiment 5 tons/acre). However, larger applications of compost will not do any harm. In fact, more residual organic matter will be produced and the soil will benefit from the same for following crops. Of course, different optimal applications will probably be found for other crops.

Probably the most important observation made is that the smaller amounts of NPK in compost produce as good, or even better, results than the larger amounts of NPK in the fertilizer. In fact, 40 lbs/acre N, 50 lbs/acre P, and 23 lbs/acre K in compost produced a markedly better over-all effect than the optimum fertilizer application (800 lbs/acre) of 48 lbs N, 80 lbs P, 32 lbs K in the 68-4 fertilizer group.

In the compost group, the application of 1, 5, and 8 tons/acre of compost produced an about equal proportion between the weight of the roots and that of the green parts. With the application of 12 tons of compost per acre, a larger proportion of root weight and length than of green parts was produced. When 17 tons/acre of compost were applied, as a mulch, the root proportion was smaller than the weight of the green parts. When the same amount of compost (17 tons/acre) was put in the planting hole, the root proportion was equal to the green parts. The use of the compost directly in the planting hole also brought about an increased development of large legume nodules (by weight, 800 mgm. as against 650 mgm). Little difference in nodule development was observed as between applications of 8 or 12 tons per acre of the compost.

One ton of compost, with 8 lbs. of N, 10 lbs. of P and 4.6 lbs. of K, yielded as much or even better than 1000 lbs. of 6-8-4 fertilizer, with 60 lbs. N, 80 lbs. P, 40 lbs. K, and this in spite of the fact that the NPK in compost is much less "available" than in the fertilizer. The term "available" as developed in connection with the fertilizer theory does not apply to compost.

CONCLUSIONS:

On peas, 5 tons of compost per acre will produce an optimum effect if broadcast or used as a side-dressing near the plants and worked into the surface soil (upper four inches). If compost is used as a mulch, a somewhat larger amount will be needed. Compost used directly in the planting hole will increase the development of roots, which might be desirable in dry climates, in the case of heavy soils and if seeds are sown late in the season.

CHAPTER IV

Raw Materials Useful for Composting

The content of chemical components valuable as fertilizer, ease of handling and transport, as well as structure and moisture content are decisive factors for the compost manufacturer in determining to use raw materials for composting.

From the biochemical point of view, the adaptability of these materials to compost fermentation is important. One should know whether the materials will decompose and in what way, the specific structure, whether aeration is necessary or not, the moisture content and bacterial action needed for producing the best results. Obtaining such information is the task of the Research and Field Laboratory.

Finally, the finished compost should lend itself easily to bagging and shipping. Its agricultural fertilizer value should be determined in order to prepare sales promotion and proper distribution to the customers, bulk to farmers and retail to home and gardens and the nursery trade. All these factors, processes and problems should be well-known and understood before a decision is made to use any specific raw material for composting. The Biochemical Research Laboratory has done a great deal of valuable research and has accumulated a wealth of knowledge and information in this field. The following raw materials have already been investigated in regard to their value and usefulness:

Bagasse and Filter Mud	Municipal Garbage
Cannery Wastes	Peat and Muck
Coffee Grounds	Sawdust
Coffee Wastes from Plantations	Sludge
Cocoa Wastes	Straw
Corncobs	Synthane Chips (a palstic)

Cotton Gin Wastes	Tea Wastes from extraction and Plantations
Hair	Water Hyacinths and Seaweed
Manure (all kinds)	Wool Wastes
Meat Processing Wastes	

The following lists some of the major chemicals contained in the raw materials investigated thus far. Since nitrogen is the most important factor in the final analysis of the product and the moisture content is essential in handling, these analyses should be made at once. In addition the phosphate, potash and trace mineral contents are determined. The methods used for the analyses are standard A.O.A.C.

For practical purposes the following descriptions of differences and variations and their significance are necessary and useful.

A. Materials low in nitrogen, i.e., below 0.7% nitrogen. Such materials ferment slowly due to the time needed for the transformation of carbon until a favorable carbon-nitrogen ratio is reached. During this process a high percentage of carbon compounds (organic matter) is lost. It is therefore preferable to use mixtures of higher nitrogen rating.

The price of low nitrogen materials and the cost of transportation should be very low. In fact, if they can be obtained and delivered free of cost it is preferable since they serve more or less only as "fillers". One exception may be large sawdust piles near lumber mills. Bagasse, paper, plastic chips, rice hulls, street cleanings and wood bark are examples of low nitrogen raw materials.

B. Materials of moderate nitrogen content, from 0.7 to 1.5% with an average of 1.0%. These materials make up the bulk of any composting process. They can be used without nitrogen correction and will produce a good compost with an average of 1.0% nitrogen. Such composts can be classified as "soil builders". Such average nitrogen bearing materials are cachaza, cocoa tankage, coffee grounds, cotton mote cleanings, garbage, grape pomace, grape seeds, licorice root, wattle bark, olive seeds, ramie wastes, some sewage sludges, also tea leaves. The price of such materials delivered to the plant should be very low.

17

C. Materials high in nitrogen content, from 1.5% upward. These are valuable in improving the quality of the finished product. The more of these which can be obtained, the better the compost will be. The amount to be used in mixture with low nitrogen materials will be controlled by the cost price rather than by the need for nitrogen. If enough of these materials can be used to get a fertilizer formula of better than 2.0% so that the end product can be sold as "fertilizer" rather than as a compost or soil builder, a higher price can be paid for it since the finished product can be sold for a higher price. In general, their value can be determined by the nitrogen content.

Organic materials high in nitrogen content are: bean pomace, blood meal, castor bean pomace, chrysanthemum waste, some filter mud on a dry basis, fish wastes, hoof and horn, kenaf wastes, pyrethrum waste, sheep wool, slaughterhouse wastes, tobacco dust and tomato pomace. The high nitrogen bearing materials are the most profitable materials in a compost mixture. If the bulk fraction (described under B) is obtained at a very low cost, more of these addenda can be used.

D. Manures, these are very important and necessary addenda to compost mixtures because of enzymes and growth factors which cannot be obtained in any other way. If manures are not available or the cost is too high, a manure solution could be used consisting of one part of manure, suspended in 100,000 parts of water and added to the compost mixture.

Ordinarily 10 to 20% manure should be added to the total compost mixture. The value of manure is partly based on its nitrogen and phosphate content and partly on its intrinsic value in creating a favorable reaction in the fermentation piles.

The very best manures are cow and pigeon. The following are classified as good: steer manure, poultry (high in nitrogen content, but less easily fermented than cow, goose, duck, sheep, or mixed stockyard manures). Pig manure is of medium value and horse and mule of low.

The price is generally determined not by the "value" of the manure, but by the availability in any given area.

18

E. Soils are an essential part of the compost mixture. Usually between 10 and 20% is used. A daily Production of 100 tons of compost would require 10 - 20 tons of soil. In some areas soil is difficult to obtain. In such cases, the tailings from screening the finished compost, or the finished compost itself, can be recirculated at the rate of 10 to 20% of the mixture. The value of the tailings is the actual cost price at the point in the process where they are removed. It is obvious that soil is cheaper. In practice soil can be stockpiled and bought at the rate of a year's supply at one time, from building excavations, road constructions, drainage work, etc.

Good humus soils (topsoils) are preferable to subsoils. Stockpiled and aged, as well as aerated subsoils are better than fresh subsoils. Eventually the growth of legumes or other green manure crops on stockpiled soils should be given consideration as a means of improving them.

Not all soils are suitable for composting. Sand, for instance, is very poor, chemically as well as structurally, and does not hold moisture. Very heavy, poor, clay and adobe soils have too fine a structure to be of value since they fill all the pores and form a film impenetrable by air. The best soils are the medium heavy humus soils.

F. Peat. In general this material is not used for composting as it takes such a long time to bacterialize. Some peats are high in nitrogen which is deceptive for the compost manufacturer. He should not be led astray by this. Then, too, peats are usually rather expensive.

G. Mucksoils. These can be used instead of humus and topsoils. They are high in organic matter and some are higher than 1.5% in nitrogen Acid muck soils should undergo an aeration process before being used for composting. Neutral or alkaline muck soils can be used as is, and are excellent materials.. However, the moisture content should be taken into consideration.

H. Cannery Wastes, from fruit and vegetable canneries. These materials are usually wet and pulpy, often containing 80 to 90% moisture, which is a handicap. A special process of decomposition in pits has been designed. The muck or dry residue from these pits is

19

excellent. The excess moisture should be compensated for if the pulpy wastes are added to the original compost mixture. If pre-fermented before use, a certain price for the material is justified; otherwise it should be delivered to the plant free of cost or for a low trucking fee.

Some cannery wastes may contain a high residue of insecticides and pesticides which will have a delaying effect on the fermentation process. It has been observed that tomato skins rotted much more slowly than the inner pulp for this reason. Wastes which contain arsenic or large amounts of copper should not be used.

I. **Pomace of any fruit, or press residue from castor beans or olives** can very well be fermented and should be evaluated according to the nitrogen content. Castor bean pomace is excellent with a 4% nitrogen content; others are low in nitrogen and will do better in mixtures.

J. **Sawdust.** This is a bulk filler and adds nothing but bulk and nitrogen free organic matter. The difference between hardwood and softwood sawdust is negligible. However, hardwood sawdust makes a better compost and ferments more easily. In general, a total compost mixture should not contain more than 30% sawdust. Large chunks and wood chips are not suitable for composting but occasionally will be attacked by bacterial action. They are eventually removed in screening. Wood ashes are high in potash, alkaline, and should be used only in high dilution (1 to 5% of the mixture).

K. **Sewage Sludge** (aerated). If used in a general compost mixture not more than 20 to 30% of the total should be added. There is such a difference in sludges that each needs to be investigated separately.

L. **Industrial Wastes.** Some of the wastes, such as tea leaves, coffee grounds, cocoa tankage, filter mud from sugar cane or beet sugar processing, beet tops from sugar beets, are all excellent source materials. Bagasse and fiber processing wastes are low in value and follow the pattern outlined for sawdust. All of these should be valued on the basis of their nitrogen content.

A special role is played by pulpy wastes from fibrous plants and from cocoa and coffee beans at the source. These make excellent materials for composting but are usually not available except in plantation areas. Some of these wastes are valuable for their content in enzymes and vitamins (B complex and B 12 and other extracts for raw materials for cortison).

M. Slaughterhouse and Rendering Plant Offal. Blood, meat scraps and tankage are high in nitrogen content and very valuable. Bones are high in phosphate, hoof and horn meal high in nitrogen. While blood and tankage can be used as they are in amounts up to 30% of the general mixture, hoof and horn should be shredded or milled before use.

Paunch manure is somewhat resistant to decomposition if not aerated and mixed with other materials. A general compost mixture should not contain more than 20% of paunch manure. Its nitrogen value is not too high.

N. Fish Wastes. These materials are medium in nitrogen content and high in phosphates and are most desirable additions to compost mixtures. 30% of the total mixture in the form of fish wastes can be easily digested in fermentation piles. Forty percent (50% with 30% soil) should be the upper limit. This is the situation with fresh fish wastes, heads, tails, entrails. Processing wastes and dry materials can also be used to advantage.

O. Corn Cobs. These make a good material, but should be well ground and soaked in water prior to mixing with other materials.

P. Cotton Wastes. The nitrogen content determines the value. If the nitrogen content is too low, these materials are about like sawdust. High nitrogen content cotton wastes will react like garbage or corncobs. They should be well moistened.

Q. Filter Mud (Cachaza). Special instructions can be obtained upon request.

R. Coal Ashes. Resist bacterial activation and should be avoided if possible. In many places they are mixed with the garbage during winter weather. In such cases the soil and if possible the manure content should be increased.

21

S. Paper. Up to 30% (by weight) of paper in garbage can easily be broken down. It should be shredded, which is usually done in the grinding or macerating and mixing process. Paper is very beneficial in loosening a tight and sticky mixture. The paper fraction can be increased in such cases, also where the moisture content of the mixture is high. Piles containing a large amount of paper heat up more than others and need more care in regard to water supply and moisture control.

Ten percent paper in the garbage fraction offers no difficulty. In our process printer's ink does not interfere with the bacterial action. Chemically, paper plays the same role as sawdust or cellulose material. The nitrogen content is about zero and its fertilizer value minimal. Its true value lies in the loosening property and it can be important in mixtures containing much cannery waste or other pulpy materials.

There are, perhaps, other specific materials which would be useful in composting; however, those described pretty well cover all parts of the field. Certainly a variety of combinations and mixtures can be made up from these various sources in accordance with their availability in specific instances.

CHAPTER V

Composting and Soil Problems in Southern, Warm and Dry Climates

Dry, warm climates present quite a few problems which are different from those encountered in moderate and Northern climates. Drought counteracts bacterial action and earthworm growth, for all the microlife of the soil needs moisture. Then too, warm soils, which heat up easily, weather away very quickly under prolonged, intensive sun radiation. Under such conditions, even granite rock gets brittle and swiftly weathers into a sandy, gravelly soil. Such granite decomposition soil, in the South and Southwest, is very fertile as long as the rains and natural moisture support plant growth. Without moisture it can remain desert. Chemical analysis reveals a high percentage of available minerals, potassium and phosphate, for instance, and less calcium. If there is some organic matter present, then with the rains, a high nitrate content might be found. But without organic matter, these soils wash out very easily and lose their high chemical rating.

The washed-out products together with clay and soluble humic acids are then deposited in the valley bottoms. There they may cake into a hardpan, the socalled adobe soils of Southern California, which make excellent material for adobe building bricks, but are exasperating to cultivate. These soils fall thus under two extreme categories, the hard caked variety, just mentioned, and the very loose crumbly type on the slopes. The latter, with alternate increasing washing and burning, become unfertile, eroded and gullied unless there is a plant cover.

The amount of clay present determines whether a soil will hold its own or lose out. Clay is a colloid which absorbs and retains moisture. When drought sets in, the clay soils hold out longer, thus enabling soil

23

life to establish itself over a longer period and to form humus. Once humus forms and is not lost again, the situation is fairly secure. The soils with humus hold and lend themselves to successful cultivation. As long as there is moisture, rain, even mist, and/or irrigation these soils will be extremely fertile. However, humus is broken down under intensive sun radiation and conditions of drought. Shading with cover crops or mulching is extremely important. One can see the most beautiful, crumbly soil, rich in organic matter, underneath the trees in an orange grove as far as their shade reaches and the fallen leaves mulch the soil. Only one layer of leaves is enough to do the job. A few inches away, on the unprotected interspace between the rows, one can see hard crusted surfaces, sometimes so hard that they will not even absorb the irrigation water. Simply raking the leaves out from under these very same trees (where the soil is sheltered enough anyhow) and spreading them over the interspaces would bring about an excellent soil condition there. However, the present orange grower thinks this would require too much labor and spends more time and money in cultivating the crust, or, more recently, in applying light oils to prevent weed growth, or in other operations.

It is claimed that this practice of oiling does not crust the soil. The writer saw oiled soils which were not crusted and others which were. The fundamental difference seems to be that the crusted ones had been mineralized before oiling. The oiling then increased the crust, while in other soils a sufficient amount of manure and compost and soil life too was present so that the soils reacted favorably and have been able to stand the oiling. Nobody can predict how long this will be so.

Recently, the writer had the opportunity of studying an interesting example of these Southwestern soils. Granite decomposition soils as well as Mojave Dester soils can be relatively rich in available minerals. Desert dust is rich in potassium. Soil analyses can be deceiving to the inexperienced. The owner of the soil analyzed may even point to a neighbor's place where - with sufficient irrigation - lush alfalfa grows. On the basis of the high mineral content of his own soil, he tries the same and nothing grows. The difference is that the neighbor's soil contains a clay as well as decomposition sand and has a clay pan nine inches below the surface while his soils contain no clay and are made up only of decomposition soils and sand, three to four feet in depth. In the latter case all moisture is lost, seeping downward and evaporating quickly, because sandy, granite soils heat up very quickly, while clay and humus are "cold" and thus lose less moisture. The sandy decomposition soil needs more irrigation, but it too will be lost. In order to succeed, such soils would need clay plus humus, plus shade, plus mulch.

However, if these soils are constantly watered, and do not dry out, then a natural balance between precipitation and evaporation can be maintained and a most beautiful humus soil will develop, even without the addition of organic matter, compost or manure, chiefly by means of the microlife naturally present. It is then up to man to maintain the fertility and ideal conditions by cover crops, avoiding excessive irrigation which would cake the soil, and by proper crop rotations which balance exhausting crops (corn, cabbage, beets) with restoring crops (legumes, or plants and trees which shade the soil). In summing up, one can say that the control of moisture, shade and humus decides whether such soils will hold or be lost.

This behavior of soils can give us leads for the development of proper composting methods in a hot, dry climate. The most important rule: Never let your compost pile dry out. Eighty per cent of all the compost piles which the writer saw in California in recent weeks were too dry. When such material is spread on a dry or caked soil it will not be absorbed, even when ploughed or harrowed under. It will lie inert, and have little value until the rain comes. Much valuable time is lost. It is this time which counts in the improvement of the soil and not the time which is spent in attending properly to compost practices. The composting which is done right will pay.

Bulky material such as dried-out tall weeds, the leaves of orange, avocado, and eucalyptus trees, even sage, can be composted if shredded. Many of the compost heaps contained these materials but they were dry with lots of air spaces and dry soil in between. No fermentation could take place. It is best, therefore, to place a compost heap under the shelter of trees to help against its drying out. (Contrary to general belief, eucalyptus trees do no harm when used for wind protection. If they grow too tall, top them and sever the roots which tend to spread out too far into the orchard or field.) If there is no natural shelter available, a simple shed can be built to take its place. Set a few upright posts, these need not be high, just tall

25

enough to enable you to work the heap easily. Then use some perforated material as a cover (tin, for instance) which will give enough shade yet let the rain through. Slant this roof so that the rain can still run into the heap. A chicken wire screen covered with large leaves, Spanish moss or weeds also makes an adequate shelter. If you have irrigation at hand, lead a ditch or pipe near the heap. Build the heap up in a pit, about two feet in depth and let the water soak through from underneath and around the heap. In the first stages you can let the water run on the material. It is very important that all dry material be wet down thoroughly while building your heap. After shredding, mix the organic material with earth, and never forget to cover the whole heap as soon as it is completed. If you do not have enough earth for this cover then use leaves, old sacking, plants, anything which does not seal it up yet still provides a skin for it. Under the warm conditions prevalent there in the South, fermentation will be much more rapid than in the North. Well built heaps are often ready after two months. And the writer has seen heaps in fine condition at considerably less than two months old, where the pit system was used with the addition of earthworms.

A completely rotted compost should be used up quickly. Of course, as long as it is kept spongy-moist and covered, it will keep for many months, maybe more than a year. But if it is allowed to dry out it will lose a great deal, particularly nitrogen. It should be worked into the surface right with the planting of the crops.

A half-rotted compost of coarse material is best for mulching on top of the soil, covering the interspaces between plants or trees. It is only necessary to mulch between the rows in groves of full-grown orange and avocado trees, or of young walnut trees. For most young trees mulching nearer to the trunk is advisable, unless one uses cover crops. But a ring, one to two feet in width, should be left free around the trunk. Sometimes one sees the compost or mulch piling up against the tree trunks. When this happens, the bark is covered and rots, is given over to infections and pests which hide underneath the loosened bark. When the area around the trees builds up too high through mulching, it should be leveled out again. One often sees trees in gardens which are too deeply "buried". If you dig out such a tree you will see old roots growing horizontally and downward, while the young roots grow upward. This might cause disturbances in the circulation of the sap, resulting in the gum bleeding of apricots, almonds and peaches.

In connection with the humus management, the right handling of irrigation is also important. Basin irrigation, where the water stands for a long time until it soaks in, crusts the soil and produces a hardpan underneath. Caked surfaces resist irrigation and lead to water losses. Mineralized soils, without absorbing humus, also behave in this way. The cultivated surface is best for irrigation; so is the furrow system. Here, one allows just enough water to run down the furrow to reach the other end. So often one hears complaints about the lack of water and then sees people wasting great quantities through wrong practices on caked soils. The most ideal conditions for absorption are present just after a mulch cover has been worked in or removed. After irrigating orchards or gardens, one might mulch again to hold the moisture. This is easily done in small gardens, but large areas present difficulties. Some day a mulching machine might be constructed which drops the material like a manure spreader and rakes the mulch into windrows, then spreads it from the rows out over the land. This machine might be a kind of combination of the side delivery and dump rakes, i.e. reversing the windrow action. So far as we know, such machines have not yet been experimented with, but it is to be hoped that somebody with initiative and mechanical, as well as organic, insight may develop adequate machinery for large scale operations.

Agriculture in extreme climatic conditions such as those of Florida or California tends towards onesidedness. The diversified pattern has been lost. The fact that one can make a living by growing a few specialized crops on small acreages has produced such phenomena as the "farmer" or "orchadist" who has only beans, or artichokes, or lettuce, or oranges, or avocados. These partial or complete monocultures draw upon the soil resources in a onesided way. Mineral and trace mineral deficiencies occur, plants gradually get weaker, a very specialized, monotonous microflora develops and the farmer is surprised that plant diseases and pests increase. In order to survive, he takes the labor and expense of spraying insecticides in his stride. When balancing organic methods are mentioned the objection is raised that they are too complicated and involve too much labor. Old orange growers tell us that in their youth, maybe 40 or 50 years ago, no pests or diseases were encountered. In those times, manure and mulching was the order of the day. They changed to modern methods. It is so easy to call up the nearest airport and have a plane come and spray insecticides. The changes of quality in foods and soils which result are not taken into consideration. The general public is not yet quality and health conscious. People will judge by size, packaging and labeling.

Recently a "quick die disease" in oranges worries the California growers. One can observe the older growers putting all the manure and mulch they can scrape up back onto the orchards in order to save the situation. Others believe in oiling. Few as yet have the courage to try organic methods. Indeed, it is difficult to introduce a balanced, diversified system in a situation as onesided as a citrus grove or a vegetable farm which produces only one or two kinds of crops. It may be necessary to compromise, for example, by using as many different kinds of materials as possible for compost (garbage, leaves, weeds, etc.) in order to add something "different" to the soil.

In California, alfalfa hay is considered by the old timers to be the ideal mulch for fruit trees. One of these men told us that he used it as long as it was cheap and never had any trouble with his trees. Nowadays it is too expensive, but organic composting methods could take its place. This same man told us that were one of his groves to die off, he would plant alfalfa to rejuvenate the soil and sell the hay for mulch. He felt that this would be more profitable. Twenty years ago, it took 125 oranges to fill a box; now it takes 185 to 225 to fill the same size box. The harvest of a successful grower, two decades ago, was 1100 boxes per acre. From the same land, he now harvests 800 boxes.

Were I in possession of an orange grove or a truck garden monoculture, I would try to grab any and every possible organic material from outside, plus small amounts of manure, and apply it to my land. As a matter of fact, I observed in many of the small one to three acre lots and gardens that, with the lush and fast growth of trees in these Southern climates and sufficient irrigation, such large masses of organic matter can be grown that they more than suffice for the fertilizer needs of the land. In one or two gardens of mixed fruit trees and vegetables, a two-year supply of compost was collected in one year. In fact, it was necessary to advise the owner to step his production down a bit. In this particular situation, it will come true that compost will be a natural cash crop.

28

It is the small lot which can be intensively developed, while with increasing size the problems grow too. This again should stimulate consideration of a possible future development: subdivision in mixed cultures which can be handled by one man instead of large monocultures. Pests dwindle away in such small controlled areas. California, at present, is considered by many to be the new frontier. She has a weekly influx of thousands of people who seek a better future in a favorable climate. Large monocultures will not be available to nourish and supply such an increasing population with work. The small, diversified grower could make a living and survive. The trend in population, the soil and health conditions, favor the natural, organic, self-sustaining unit. These alone will be able to absorb the increasing population and make possible its survival.

CHAPTER VI

A General Review of the Problem of Composting
(Composting Again — small and large)

Composting of organic wastes on the farm or in the garden is an old art, probably as old as man's tilling of the soil. Old-fashioned composting has been practiced right up to our own times. But it remained for the modern agriculturist, the farmer who has to work under the pressure of time, wage scales and scientific fertilizing, to find out that these methods are too expensive. Hence he tried to replace composting in one way or another. The good gardener never entirely lost sight of it.

The scientific age with its teaching of mineral deficiencies and fertilizers, the so-called "Liebig Age", looked down on composting for a while, only to discover in time that without organic matter in the soil, minerals alone will not solve the problem. So composting has become a new skill to be learned. And at this time, under the scrutiny of science, progress has been made in many regards, and we are now on the threshold of a new age—The Organic Age—when composting becomes a science—and an industry.

The rules of small composting in the home garden or on the farm have been so well publicized in recent years that we need only mention that the building of a good pile is a matter of skill. Practice has also shown that when scientific procedures are used, the pile can rot down to a fine humus in half, or less, than the time that was needed when following the old well-trodden path.

The purpose is in any case to get humus and to incorporate this humus into the soil. The points of view about humus and compost have of necessity changed a bit, and are still changing.

Humus, as it derives from decaying plants, refuse, and manure, is not fertilizer in the sense the word is used nowadays. One argument from the orthodox "plant food" and fertilizer camp is that there is not enough composthumus available to satisfy the needs of a modern, intensified agriculture. This is, regrettably, true.

It should be realized that not all decayed or broken down organic matter is compost, in the true sense of the word. Compost is more correctly defined as a completely digested, earthy matter having the properties and structure of humus. At present many products are called "compost" which have not yet reached the humus state or have gone too far, resembling good soil but lacking organic matter. There are many different kinds of compost with quite different fertilizer values and effects upon the soil. Proper composting is more a matter of science and proper technology.

It is generally acknowledged that a farmer should use 10 tons of farmyard manure per acre, in a proper crop rotation, in order to maintain his land in a good state of fertility. There is not enough manure available to carry out this rule of experience, especially in areas where there is no livestock and where the tractor has replaced the horse. The small farmer of the underdeveloped areas of the world, who has to make a living from a few acres, has it easier in this regard than the big operator who farms hundreds or even thousands of acres intensively.

Evidently humus helps to "hold the soil". It absorbs and retains moisture, so that humus soils are more resistant to drought and also can take better care of excessive moisture. Humus soils are of crumbly structure due to the microlife which develops in them. They hold the mineral supplies in an available state, ready for plant growth. Virgin soils used to have 4 to 6% organic matter; today the U.S. average for agricultural soils is 1.5% organic matter. There are many soils, especially on truck farms, below this average which, by the way, is already the critical level, indicating a breakdown of the soil. Good soils have better than two percent organic matter, but these become rarer and rarer.

Composting is not the job of a privileged class of enthusiasts and organic gardeners, called by some people "organic faddists", but rather it is a *must* unless we are to lose out with our soils, adding complete depletion to erosion.

At present the use of organic wastes in the form of compost is looked upon as a soil conditioner and as food for the microorganisms

31

in the soil, rather than as fertilizer. How important it is to "feed" the soil may be realized if it is known that an acre of good soil contains, to a depth of about 7", up to 1,500 pounds of soil bacteria and 750 pounds of fungi and tiny animal organisms.

In order to simplify matters, and to economize, the suggestion of green manuring has been made, also the use of so-called sheet composting. No doubt green manuring adds a considerable amount of bulk organic materials, ranging between 5-20 tons per acre. Heavy, wet soils, however, do not take well to plowed-under green manures; and all soils are tied down by them for a considerable length of time. In fact, observation has shown that green manuring consumes microlife, and temporary nitrogen deficiencies are caused, lasting until the green matter is decomposed. Sheet composting means that the crude, sour wastes like garbage are spread over the fields and turned under as they are. This also ties down soil life and nitrogen for a while, and creates different structural conditions until the wastes are decomposed. Should there be diseases, fungi, insect eggs, larvae in the waste, there is no better way to spread them than by sheet composting. So we come back to composting as a must; applying decomposed, humified matter which is readily available. Compost comes in handy where there is not enough manure available, especially for truck farming, farms without livestock or not enough, monocultures and tropical cultures. Its value, besides the fertilizer elements contained in it, lies in the structural improvement of the soil, and its water holding and moisture absorption qualities. Even manure is improved by composting. Sewage sludge should by all means be composted in order to destroy existing intestinal organisms and the after effects of anaerobic fermentation. As a productive waste disposal, composting ranks at the top. If properly done it is also the most sanitary method of waste disposal.

Industrial composting methods have developed in the past decades along various lines, using garbage, stockyard manure, industrial wastes. Some of these methods were successful, others not. Most proved to be too expensive due to costly machinery, and handling, i.e. labor, involved. Almost every country has had some experience with one or another process. If one starts to introduce a new process, as this writer did, one is confronted with history and is told: Yes, we see it is a good thing to compost wastes, we need organic matter for our soils, but this man and that man and that other man have all tried it and found it too expensive and had to give up again. Some have developed a product which sells for a premium, it would be more accurate to say, sells as a luxury item, considering the amount of money a farmer can afford to spend per acre by way of organic fertilizer.

In order to solve the problem of the composting of industrial and urban wastes, two questions have to be answered: (a) Can it be done, taking care of a large volume of material per day? (b) Can it be done economically? Many a process which looks good when 5 or 10 tons a day are produced becomes too troublesome and expensive when 50 to 100 tons a day have to be treated.

It is not the purpose of this article to give a critical analysis of composting processes of the past or present. A few names, which have been, or are, in the public eye, are mentioned by way of a historical record. There is, for instance, Sir Albert Howard's Indore Method, mixing and piling source materials. The fermentation is left more or less to nature and chance, though the conditions of an "ideal" compost pile have been very well developed. Frequent turning of the pile produces the best effect. This, alas, reduces the usefulness of the method for large scale operations.

From France and Italy there have come methods which make use of bins, aerated bins, fermentation cells. When tried in this country, they proved to be cumbersome and not sufficiently up to economic standards, and so had to be given up. Verdier's and Baccari's processes may be mentioned at this point. There is quite an extensive patent literature on digesters in which the materials to be composted are mechanically transported and pass through a digester which maintain most favorable aerobic conditions of fermentation. Other digesters maintain cells, bins under anaerobic fermentation. In some of these instances starters or activators or cultures are added. The best known processes are connected with such names as Earp-Thomas, Frazer, American Composter Company. All these methods will produce compost. Whether they can handle garbage at the rate of 100 or more tons a day and do this economically is another question. In introducing his method, the writer of this article has had all the others quoted against him in being told it cannot be done economically.

It is evident that frequent turning of the piles is a handicap. Likewise it is evident that unsorted garbage containing trash, glass,

33

tin cans, china, is a serious handicap. The Pfeiffer Process suggested the use of sorted garbage obtained by a separate so-called two-can collection. In municipalities where such separate collection is practiced, the problem is simple. The plant in Oakland, operated by the Compost Corporation of America, used a separating unit with a sorting belt, which was already in existence. This unit had been built and used previously by the Oakland Scavenger Company for other purposes. The operation delivered a "clean" garbage. A magnetic pulley had been added, which removed tin cans and iron objects successfuly. The initial expense for a sorting unit is high, and it will pay only in large-scale operations of 60 to 100 or more tons per day. It would, however, be prohibitive for small town operation of 10, 20 or 30 tons per day. No doubt in due time an experienced engineering company will come up with the solution, in terms of completely mechanical sorting.

....

The fact that the last described methods work without using bacterial starters is occasionally used as an argument against the Pfeiffer Method which uses a starter. Of course, one can obtain compost without the use of a bacterial starter. Composting without the addition of a starter is found all through history. In such cases the organic work is done by various organisms which just happen to be present in the waste materials. It would be ignoring scientific progress if we were not to investigate the use of bacteria cultures especially suited to the purpose. Those familiar with the science and practice of fermentation in industry know very well that controlled fermentation, operating with specific yeasts or cultures, works economically and successfully to produce, in engineered processes, end materials excelling in quality. No good wine, beer, baker's yeast, antibiotics, lactic acid, or vinegar could be made were it not for the fact that definite, well-known cultures are used as "seed".

The author undertook a painstaking study of all suitable microorganisms before he arrived at his solution of the problem of a bacterial starter. In it, selected organisms, concerning which all laboratory, microbiological and biochemical data are known, are used in a well determined ratio in order to produce a directed fermentation under controlled conditions. Specific mixtures for definite purposes have been developed, for instance, for wet garbage,

34

for garbage containing a lot of paper, for industrial wastes, etc. The Biochemical Research Laboratory, Spring Valley, N. Y., has developed bacteria in a concentrated form (known as B.D. Compost Starter) for the composting of various industrial wastes. A list of materials which can be treated successfully with the B.D. Starter includes garbage, bagasse, filter mud or cachaza, cannery waste, coffee grounds, corn cobs, cotton gin wastes, cocoa tankage, tea leaves, ramie and kenaf wastes, flax shives, rendering offal, night soil, municipal garbage, sludge, straw, synthane chips (a plastic) and wool wastes.

The laboratory is in a position to study specific problems submitted to it. Suggestions for the proper formula for a particular fermentation can then be formulated as a result of such studies. It will usually take from two to three months to conduct an extensive laboratory test. In connection with any special problem, please state the nature of the material and send a five pound sample of it for testing purposes.

The B.D. Compost Starter will decompose only organic materials. It is not harmful to living organisms. It does not attack living plants. It was inhaled as dust by workers and was injected into test animals. No ill effects or infections were observed. The writer's dog loves to eat it. It should be remembered that many of the organisms used are related to the large family of organisms which produce antibiotics and vitamin B12. There is also no fly and mosquito breeding on the compost piles, partially due to the fact that flies thrive on putrefactive material but do not lay eggs in good earthy compost material. We have never seen rats or mice around the piles. It should also be noted that bacterial action may be slowed down by cold and frost but will not be otherwise impaired.

The laboratory has worked out and suggested various mixtures of organic source materials in order to obtain the conditions most favorable to fermentation. The aim is a carbon to nitrogen ratio of between 20:1 and 11:1 (optimal effect). That is to say, the whole equipment of the analytical laboratory has been employed in order to get an understanding of all phases of the process. It can be stated that under such conditions, composting is no longer a chance proposition, but rather an exact, quantitative science. Laboratory tests and calculations made in my laboratory differed by only 0.2% from the analysis of the product which was obtained many months later in the industrial application of the process. We refer to the final chemical formula of the finished product. Only if composting has been made in a quantitative science is it possible to work out the economic problems involved in the processing, and to determine both the costs and the exact procedures to be followed. At present, all raw source materials are tested with suitable strains of bacteria, the C:N relationship is investigated and the actual procedure for large-scale operations is tried out in the laboratory before erecting a plant. The process in the plant is constantly checked by analyses and the fermentation is controlled. All this enables the manufacturer to avoid costly operations such as frequent turning of the piles, long waiting periods for the piles to be completely rotted; also avoided are undesirable odors, and losses of organic matter due to oxidation (burning up).

We mention here by way of illustration the analysis of a typical compost, an organic fertilizer made from city wastes according to our process. Mixture of raw materials used:

Garbage 70%   Soil 10%   Manure 20%

This particular garbage mixture contained one third of its weight in paper and cardboard wastes, i.e., 23% of the total compost mixture was paper. It was average city garbage collected in late winter. The material was completely digested and air-dried to a moisture content of 20%. Swill produces a considerably higher analysis than the one given below as "average". Organic matter was 27%, the balance mineral compounds. The degree of fermentation and the content of organic matter and nitrogen make a great difference between Starter and non-starter made composts. Machine macerated composts in Europe are usually low in organic matter (below 18%). Here is the great difference—in the quality—between B.D. Starter produced composts and other products. The availability of minerals is also quite different. Therefore much smaller quantities of the B.D. Starter treated compost, especially in its dry form, need to be used. We deal here with an organic concentrate.

Analysis of the afore mentioned product:

Total Nitrogen: 1.0% on the average. During spring and summer when more vegetables are used, the nitrogen climbed to 1.4%. Changing the formula by including poultry manure, high protein cannery wastes such as pea vines, asparagus wastes, fish, and meat wastes, tends to increase the nitrogen content up to 2.0%. Nitrate

36

nitrogen: 0.3%, ammonia nitrogen: 0.001%. The bacterial action is directed in such a way that organic nitrogen incrases but free ammonia is avoided. The addition of nitrogen rich source materials can raise the total nitrogen to 3% with an upper limit of 4.0%. Most nitrogen is present as organic nitrogen.

Total Phosphates: 2.0% on the average, with a fluctuation between 1.6 and 2.4%, the latter resulting from the use of fish wastes. Bone can increase the phosphate content to 3.0%.

Available Phosphate: 1.8% on the average, with a fluctuation not below 1.5%. This high availability is due to bacterial action and is a typical feature of treatment with the B.D. Compost Starter.

Total Potash: 1.0% on the average. The addition of green plant materials, leaves, wood ashes, bark and seaweed can bring the total potash up to 2.0%.

Available (water soluble) Potash: 0.5%. Continued bacterial action makes almost all potash slowly available.

Exchangeable Calcium: 0.5%.

Free Sulphates: none. Due to bacterial action the compost has a neutralizing effect on both acid and alkaline conditions, as field tests have demonstrated.

Trace Minerals: Lead, arsenic, barium, chromium were absent in our sample. However, these are not desirable.

pH: between 7.2 and 8.0. It is possible to obtain a slightly acid compost with a pH between 6.0 and 6.5 if this is required, and specific fermentation processes are used to produce this result.

Bacteria content: on the average, air-dried compost has a bacteria count between 1 and 50 billion aerobic bacteria per gram, and 10 to 300 million anaerobic bacteria per gram, using our counting method. Different counting methods give different results. Figures in literature are therefore not always comparable. Typical soil-humus bacteria make up the major fraction, if the B.D. Compost Starter is used. Actinomycetes and streptomyces are present, accounting for aboput 10 to 20% of the total content. The dehydrated compost at a moisture level of 10 to 20% still contains many bacteria that will come to life again when moistened. Dehydration below 8% causes considerable changes and cannot be recommended. The bacteria counts are made by using the plate counting method on beef agar peptone after 48 hours of incubation at 23 degrees C. for aerobic organisms, and on Brewers plates with anaerobic agar medium containing sodium thioglycollate, for anaerobic bacteria. In the carefully dehydrated compost the microorganisms are in a latent state but revive as soon as moisture is added. The re-activation takes place within 24 to 48 hours.

It is obvious that the mechanically macerated "compost" or that made without the addition of suitable organisms which crowd out the "chance" organisms, is qualitatively different from the product prescribed here. In many cases it is only a "stabilized" breakdown product and does not merit the name "compost" for it does not contain certain typical humus and soil organisms.

It is possible to enrich a compost fertilizer by using different raw materials for fermentation. It is also possible to bring this about by adding minerals (fertilizer, rockmeal, rockphosphates, etc.). Available fertilizers should be added to the finished product after fermentation. Rockmeal and rock phosphates should be added before fermentation in order to profit from their increased availability due to bacterial action.

The addition of nitrogen has been tried. Urea shows the best resuts when the total nitrogen content does not exceed 4.0%. Preferably the enriched product should contain not more than 3.0% total nitrogen. Ammonium sulphate has also been used, but it is preferable to keep the total nitrogen below 3.0% and the ammonia nitrogen not higher than 2%. These proportions have been found the most successful in field tests with growing crops. After application there is a slow but steady release of nitrogen in the soil due to bacterial action. In this regard the behavior of organic humus composts is entirely different from that of inorganic fertilizers. In comparative field tests it has been found that an enriched compost with a 4.0% total nitrogen content produced the same yield as a 16 to 20% fertilizer on a pound per pound basis of application in the case of tomatoes, corn and lettuce. In other words, the bacterialized compost makes possible an economical and lasting use of its nitrogen compounds. Ammonium nitrates are not suitable as admixtures, for in organic compounds they may lead to spontaﾞeous combustion.

Phosphates can also be added but not more than enough to bring the formula up to 4.0% if necessary. While phosphate fertilizer may be locked up in soils and become unavailable, especially in alkaline and dry soils, the continued bacterial action will provide a steady availability of phosphates in bacterialized composts.

Potash can be added in any form. It is advisable not to increase the potash content above the total level of 4.0%. The availability of potash is very much increased through bacterial action in the soil, especially in soils with a high organic content.

Agricultural limestone (a source of calcium) can be added to acid compost when large amounts of cannery wastes have been used. Magnesium can also be added. Dolomitic limestone is preferable because of its magnesium content. In general, an addition of 100 to 200 pounds per ton of compost will be sufficient. Leafmold compost and softwood sawdust compost may require the addition of larger amounts of calcium than composts made from city wastes. Lime should be omitted from composts intended for acid loving plants.

Dependent on the raw materials used and the type of fermentation applied, there are as many kinds of composts as there are brands of fertilizers. The final product should, therefore, be sold under a guaranteed analysis in order to avoid disappointment on the part of the purchaser. We have seen composts with a nitrogen or phosphate content as low as 0.4%. These were used by the customer as "compost" just as he would have used others with a nitrogen content of 2.0%. Then people report that they have tried compost with unsatisfactory results. In the interest of the user and in order that he may be able to determine the proper rate of application, it is necessary that he know exactly what he is buying rather than just "compost". One should know the ingredients and the moisture content of the compost one is using. For instance, it is generally agreed that 10 tons of farmyard manure per acre will produce a good crop and maintain the fertility of the soil. This barnyard manure contains 0.6% nitrogen and phosphate and has a moisture content of 50% or more. Dry steer manure in California tests 1.5 to 2.0% nitrogen and between 15 and 20% moisture. We can see from these two analyses that only half as much dry steer manure would be needed to equal the effect of an application of 10 tons per acre of wet barnyard manure. Even less dehydrated manufactured compost made according to the Pfeiffer Process is needed per acre.

Another objection which is often raised against the use of composts is that their structure causes difficulty in spreading. Loose, bulky, homemade compost must be spread by hand. A manure-spreader sometimes works in certain instances and fails completely in others. The right spreading method is important for large scale field application and determines whether a farmer or grower can afford to use compost. Hauling, whether by truck or freight car, is an expensive proposition these days, and the cost of transportation over long distances may make the total cost of the product prohibitive. Homemade composts, like barnyard manure, usually have a moisture content of 50 or more. It is too expensive to ship such a quantity of water. This has led to the concentration and dehydration of composts and the development of organic fertilizers made from these dehydrated materials. Dehydration by air is feasible in the southern states or wherever climatic conditions permit. Otherwise a dehydrator should be used. This too is an expensive item. A small production of 10 to 30 tons per day may not justify the investment in a dehydrator unless cheaper devices of this nature than those at present available are designed. At a higher rate of production dehydrators can very well be economic investments. Dehydrated compost (organic fertilizer) is a concentrate and will save on shipping costs. It can be applied at a much lower rate per acre, for instance 1 to 3 tons per acre, which also decreases the labor of spreading. We use a rotary-trailer lime or fertilizer spreader behind a truck and can easily spread 1 to 2 acres in half an hour, that is, at the speed the truck can travel over the field.

For home use on lawns and also on farms, some customers request a product which can be used in a fertilizer spreader of conventional design. Here a pelletized product is needed.. Dehydration and pelletizing should be done in such a way that the microorganisms in the compost and its structure are not destroyed. This is possible as we know from experience, but we have seen dehydrated manure which had lost its "wetting" capacity and was "sterilized". The pelletized product can also be used as a side dressing on row crops in ordinary fertilizer spreaders designed for such use. Side dressing has proved to be a very practical means of applying organic fertilizers. Good results have been obtained by side dressing with as little as 600 pounds per acre.

For carrying out these various operations a composting plant consists, to begin with, of a receiving platform where all incoming

40

materials are stored. A conveyor then brings the source material to a grinder or macerator. Hammer mills are not suitable for this operation. Grinding hogs or rasps or sluggers are preferable. Materials which would be abrasive are added after the grinding. A mixer is useful at this stage. Another system is to use a macerator (ball or rod mills can be used) which is followed by a second tumbler or rotary screen. The ground and mixed materials, which have also been inoculated with the B.D. Compost Starter, are then dumped in fermentation piles. A dump truck is used for transporting them to the piles. These piles, with aerobic fermentation conditions, have a "critical" size. They should not be higher than five feet or wider than twelve to fourteen feet in order to maintain aerobic conditions. They can be as long as desired. As a brewery has a brewmaster, so the composting plant needs a compost master who is thoroughly familiar with the fermentation process and who knows the behavior of all the raw materials as well as the proper mixture to be used.

The fermentation can be conducted in such a way that no obnoxious odors develop. It takes from two to four weeks, depending on climate, temperature, and moisture. The control of moisture is essential. With dry materials and in dry hot climates the addition of water may be necessary. Very wet piles rot entirely differently. A reduction of excessive moisture can be brought about by the addition of dry materials, by allowing the piles to "steam out" and by turning or shaking up piles. This last is done relatively inexpensively with a front-end loader, or better with an overhead loader.

After the fermentation is complete, the compost is air dried by flattening out the piles and disking them. The material is then moved to the dehydrator or if sufficiently dry to the screen in order to remove lumps, stones, and any objectionable bulky contents. The air dried or dehydrated material which is ready for storage and bagging should be conveyed to a storage bin under a roof or to a silo type shelter. Dry compost, below 20% moisture, does not change much, and keeps well over long periods of time. As soon as the moisture rises above 20% bacterial action begins again. If the material is sold with a guaranteed fertilizer formula, the moisture content has to be constant and dehydration is a must. Bagging of the dry material is simple and standard equipment can be used. Burlap bags draw too much moisture from the air. Therefore, polyethylene lined bags or three ply

41

bags with an inner layer of asphalt impregnated paper are preferable.

In northern cold climates the grinding and sorting machines should be under roof and in slightly heated rooms. The piles can be built in the open. If a pile freezes up, this will delay but not impair fermentation. Cases have been observed where the piles heated up and did not freeze despite low air temperature. Even a two foot snow fall melted on them. Towards the end of a fermentation a wet pile may freeze, at least on the surface. When thawing the material falls apart nicely. Certain soil bacteria continue to grow at freezing temperatures. However, in cold climate one might reckon with a somewhat slowed fermentation and therefore plan on a certain amount of stockpiling. When frozen chunks are picked up for dehydration, a hammer mill may be necessary to break them. All this means that in regions with cold winters, the operation of a composting plant will be somewhat more expensive due to extra handling and buildings than in moderate and warm climates. In areas of low rainfall all the operations can be carried on in the open and no costly buildings are necessary. In very wet climate with torrential rainfalls, such as the tropics, a shelter for the machinery and the finished product is preferable.

Finally a word regarding sales and marketing conditions. Growing out of the nature of agricultural production, sales are seasonal according to the times of the year when the farmer applies manure or fertilizer. Stockpiling of the product during the "dead" season is necessary. A compost company which processes municipal wastes arriving every day should therefore have enough operating capital to bridge over the stockpiling period for a product which is mainly sold in spring or fall.

The trade, which is used to handling so-called high grades of fertilizer concentrates with 6, 10 or more per cent of nitrogen and phosphate, sometimes wrinkles up its nose at the "low grade" formula of compost fertilizers. Here an educational program is necessary in order to demonstrate that composted organic fertilizers behave entirely differently than do the inorganic-mineral fertilizers. In the former, nitrogen is released slowly but steadily over a long period of time and therefore no losses occur. Phosphates, due to bacterial action, remain available and are not tied down. Also, the organic matter in the compost is wanted in the soil. This means that in terms of amounts where 500 pounds per acre of a "high grade" fertilizer are needed, one may use 1 to 2 tons per acre of a compost fertilizer. The price of the compost therefore has to be adjusted in accordance with what the farmer pays for his soil nutriments on a competitive price scale. Putting organic matter into the soils as the major means of maintaining fertility and structure is the chief object of compost application. The depletion of organic matter has been tremendous in recent years. Organic matter, where soil is deficient in it, cannot be replaced by mineral matter. In this lies the future of composting which, if properly applied, will also contain all the mineral nutriments plus microorganisms to enliven the soil. A compost concentrate, as outlined above, has twice to three times the fertilizer value of manure on a pound to pound basis. Since the purpose of organic fertilizers is different from that of mineral fertilizers, formulas cannot be exactly compared on a percentage basis. This should always be born in mind when establishing the price policy of a composting plant.

CHAPTER VII

The Bacterial Treatment

The breakdown and transformation of organic wastes into compost and humus take place in two steps:

A. The breakdown proper.
B. The building up phase or synthesis of new organic matter in the humus state.

Broken down organic matter is not yet humus. Therefore, a differentiation between two types of compost should be made. The first (a) is broken down organic matter which can be stabilized so that it does not produce foul odors or develop a slimy consistency. This material is frequently produced and sold as "compost". The chemical formula with regard to NPK may be identical with the other phase but the compounds are quite different. Its effect on soil and roots is different too.

In the second phase (b) we have another process where microorganisms transform the broken down (a) material into their own bodies as waste products of their metabolism which builds up real genuine humus. It is this humus in soil which produces all the benefits attributed to humus and so-called "organic matter" in soil. It is this type of humus which is produced with the Pfeiffer Process.

C. There is a third phase of compost which is frequently encountered in home garden and farm composting, namely when compost has gone too far and a great deal of organic matter has been burned up. This type of compost should properly be called "mineralized compost" since its organic matter is only between 10 and 18%. It is earthy and in itself is a good compost for practical purposes. Its chemical formula of NPK is usually low, between 0.5 and 0.8% of

44

each. It can be used, but as a commercial product it would not be rich enought to warrant its price. It contains much inert material though it may be teeming with earthworms and soil organisms.

We shall now take up the two phases of manufacturing compost:

A. The Breakdown Phase:

Many microorganisms living in air, dust, water, mud, oil, etc. attack dead organic matter and digest it. Proteins are broken down to amino acids, amines and finally to ammonia, nitrites, nitrates and even free nitrogen. Urea, uric acids and other non-protein nitrogen containing compounds are reduced to ammonia, nitrites, nitrates and free nitrogen. Carbon compounds are oxidized to carbon dioxide (aerobic) or reduced metane (anerobic). That is, CO_2 or CH_4 will escape. Sulfur containing compounds can also be oxidized to sulphate or reduced to sulfides or free Hydrogendisulfide (H_2S). If these processes do not lead to the end products of N_2, H_2O, H_2S, CO_2, but are stopped or interrupted halfway, one speaks of a stabilized product. Nitrogen is then usually contained to a high percentage as organic nitrogen. In nature this process takes place in all degrees over short or long periods of time without the aid of man. Composts of this kind existed ever since nature grew plants and animals.

Animal manures represent a special case where the breakdown phase takes place within the body. Much can be learned from this animal digestion. In the rumen, paunch or intestines of animals we have mainly anaerobic processes.

The breakdown of food is least carried on in the manure of cattle (ruminants in general) and horses, while a much greater degree of breakdown takes place in human beings and pigs. Hence the different qualities of manure, even though some manures may have a higher NPK than others but may still not be perfect humus builders in soil.

From the biochemical point of view, composts of type A from organic wastes (not from animal manure) are "vegetable manures". In other words, by means of microorganisms and enzymes these organic wastes are predigested in the compost pile as food is digested in the intestinal tract and stomach of animals in order to produce manure.

Such products may have a quick fertilizing effect in soils but no stable humus is formed, especially not on light soils or those with a gravel subsoil. Of these composts one will need more per acre and still have no lasting effects. The most outstanding research in this field has been done by Professor Springer of the Bavarian State Institute of Plant Growth about "feeding humus" and "stable humus" (Naehrhumus und Dauerhumus).

Any farmer and gardener knows the difference between fresh green manures and completely rotted, fermented or composted manures. The scientist also knows that certain rather stable digestive products give the bad odor to manures and can be carried on through the soil to growing plants. Such compounds are skatol, indol, and if sulfur is involved, the evil smelling mercaptanes. If one fertilizes a potato or cabbage patch with fresh manure and then cooks the potato or cabbage in a covered pot, one can easily smell the kind of manure which had been applied to the soil. No such odor will be present from the application of completely digested composts or humus.

Many species of microorganisms take part in the breakdown phase. Probably only a fraction of them have yet been investigated and described by science. Many are still to be identified. Furthermore, there are organisms which produce toxic products. The most typical of the latter are anaerobic bacteria of the clostridium family which lead to food poisoning, i.e., botulism and pathological organisms such as bac. tetanus (on horse manure). The growth of these organisms should be avoided by increasing the aeration of composts. Another organism of the clostridium group, which caused much trouble during World War I, was the gas gangrene producing group caused by the application of wet manures plowed under in heavy wet clay soils in the north of France and in Belgium.

The compost manufacturer is especially interested in organisms which produce ammonia, nitrites, nitrates, digest and reduce proteins (peptonization), produce carbon-dioxide, hydrogen-di-sulfide, etc. In addition, there are bacteria which produce acid and/or gas from sugars, others which hydrolize starch. In general it can be said that almost for each organic material there exists a special variety of bacteria which makes use of it. There are thermophilic bacteria, cellulose digesters and the long list of yeasts, fungi and algae which may settle in a compost pile.

It is perfectly true and correct that there are so many organisms in nature that it can always be counted on that some or other be present to start fermentation. It is necessary only to grind the materials, mix them, maintain moisture at a certain level and action will begin. This is the way nature has operated for thousands of years. Some experimenters still believe this is the only way. But we are not satisfied with that statement. It looks like defeatism and resignation. We want to make skillful use of bacteriology and direct the fermentation into the most favorable channels. Only in this way can compost manufacturing become a business, by applying the knowledge of science.

Our reasons for emphasizing the need and usefulness of a bacterial starter are:

1. The entire fermentation industry using yeasts works with very well defined organisms. Unless a specific yeast (brewers, wine, alcohol, cheese) is used the fermentation will not be efficient; the product will not be uniform; secondary reaction products with unfavorable flavors, colors and tastes may result. No one would think of baking bread with wild yeasts which are everywhere in the air and which will eventually settle on the dough by chance. It is a specific baker's yeast that is necessary. This is also the case in the production of quality wines, cheeses, etc. In all cases a specific planned fermentation is needed to produce the desired results. The condition to allow specific organisms to grow have to be known and maintained.

2. The breakdown products such as ammonia and carbon dioxide are necessary intermediate states. If their production would go too far it would mean losses. The loss of ammonia, nitrites, free nitrogen, carbon dioxide, methane, etc., would reduce the original content of nitrogen and carbon and result in a product of a lower fertilizer value, far below that of the original source materials. This is a wasteful operation. We have seen many composts of this type which contained half of the original content of nitrogen and carbon (i.e., organic matter).

By all means the breakdown whould be directed towards the maximum efficiency and conservation.

The B.D. Compost Starter, which is an essential and integral part of our process, contains a well measured and balanced mixture of the most favorable breakdown organisms, ammonifiers, nitrate formers, cellulose, sugar and starch digesters, in order to bring about the optimum results. These organisms will crowd out the undesirable ones.

3. A third and quite important reason for using a bacterial Starter is the fact that undirected fermentation by "chance" organisms can produce toxic effects. Hydrogen-disulfide, which is produced by certain organisms, is toxic and destroys many beneficial organisms, to say nothing of the foul odor of rotten eggs which emanates. Such organisms are eliminated as much as possible, since nature's chance may still add a few. Other toxins have to be avoided. Nitrites (NO_2) can have a detrimental influence upon compost fermentation, aerobic nitrate formation, and are even toxic when unchanged and absorbed by plant roots. Nitrites may also penetrate into the good water supply and render it unpalatable for human consumption. Frequently fermentation may begin to work all right but then suddenly stops. Nitrites and other toxic products may be the reason. There is always a chance that undesirable organisms including intestinal parasites and disease germs are accidentally introduced by the use of cow manures and other source materials, night soil or sludge for instance. In order to counteract these undesirable organisms before they can do any harm, it is necessary to introduce enough of the desirable organisms.

B. The Second Phase—The Building Up of Humus:

The organisms which are chiefly responsible for the transformation to humus are aerobic and facultative aerobic, sporing and non-sporing and nitrogen fixing bacteria of the azotobacter and nitrosomonas group. Actinomycetes and streptomyces play an important role. All of these are typical soil organisms. It has been our experience that these organisms grow more slowly than the ones important to Phase A. Some of them will not grow on synthetic laboratory media, nor in composts containing no soil, that is, when putrefactive fermentation prevails. We recommend therefore the addition of soil, at least 10% and preferably more, to a compost mixture in order to favor the development and survival of these valuable organisms.

Some nitrogen fixing bacteria are slow in "coming", and need at least 9 days for development, as well as very specific moisture and temperature conditions. The first stormy heated phase of fermentation has to be completed before they can perform their task. At first they will not "show". While it is theoretically possible, and has been done in laboratory experiments, to breakdown wastes in as short a time as 2 to 4 days, such fast fermentation does not allow the development of this second group of organisms. It is for this reason that we prefer the slower period of 16 to 21 days with the open pile method, since only in this way is it possible to develop the growth of typical soil and nitrogen fixing organisms, which, by the way, also makes available minerals and phosphates.

We are not allergic to digesters or tumblers. They have their place where no land for fermentation piles is available or a very fast breakdown is needed. However, the digester product itself belongs to Class A and not to Class B (humus) and may need after-ripening. Digester processes require some very specific organisms in order to be efficient, to avoid over-heating, etc.

It is in connection with Phase B that B.D. Starter treated composts differ entirely from other manufacturing processes and produce a product which is in line with slowly developed natural humus composts, leafmold and very well rotted, aged manure.

Frequent turning of mixtures in the initial phase may, in fact, speed up the Phase A breakdown but handicap the Phase B development, until complete stabilization is reached. The proper condition for Phase B we call "settling", which means that a similar process takes place to that in soils when they are plowed and/or disked. A proper "settling" of the soil is necessary before the next cultivation can be done or the seeds are sown. Structural as well as microbiological changes take place during this "settling" period.

The empirical knowledge of the experienced compost "master" is still the best safeguard to judge the conditions. Laboratory tests, moisture and temperature control, bacteriological check-ups which are time consuming, can establish the progress or delay of this phase. But the "finger tip" feeling of the experienced operator is the least expensive and quite a safe method.

The most characteristic and most recognizable symptom of a change to Phase B and its proper development is the change in color and odor. Brownish colors appear where yellow and green prevailed. The putrefactive odors, if any, in the inner layers of the compost disappear and a musty odor develops which later develops into an earthy odor. In fact that which is called "the fragrance of freshly plowed soil", especially after a rain, which is the odor of good earth, is due to certain actinomycetes. These are the organisms which produce humus.

Science has as yet very little to offer on the investigation and understanding of this phase. The Biochemical Research Laboratory was fortunate to have developed methods of research by means of which this phase can be followed and organisms favoring its development can be grown and multiplied. It is in this phase that the strength and special effects of the B.D. Starter come to the fore. Research in this field is still being carried on. We do not pretend however to know all the answers yet. From a historical point of view it is interesting and important that most studies and practical applications of B.D. treated compost and manures from 1922 to 1946 were almost exclusively done with Phase B completed material. Only the need for fast decomposition of municipal and industrial wastes which arrive daily in large amounts, necessitated the Phase A process and study. At this time, when composting seems to have become a scientifically accepted procedure to the orthodox agricultural institutions, the interest has entirely turned to Phase A and the humification proper has been neglected.

Such methods as developed by Sir Albert Howard (Indore method) and others working with forced aeration and frequent turning have usually worked only with the Phase A fermentation which is much more spectacular than the changes taking place in the later phase. Phase B will take place in nature with any compost which is kept moist (and shaded) and left to rot for a long time, but this process is very slow. It has been our task to speed up and control it so that it can be used in large scale compost manufacturing. It will definitely produce a superior product.

The beginning of Phase B which, we repeat, is the humus forming phase, is usually indicated by a drop in temperature and moisture in the pile. The latter happens only if the material had a high moisture content at the start. The moisture content should be adjusted at the time of the second phase of fermentation, or shortly before, as a drying out of the pile would completely stop the development of Phase B organisms and the effect of the Starter would be lost. This has happened here and there especially in small piles. Turning the pile and re-inoculating with the B. D. Starter would be necessary in such unfavorable cases. In large scale composting the "drop" of Phase B was less frequently observed. The ideal for Phase B is to have the pile covered with soil (1 inch layer) or paper, straw, leaves, or otherwise shaded. Wet caked materials are resistant to Phase B growth. As mentioned, in practice the problem is not as difficult as it seems to be theoretically.

Actually the pure chemical analysis of NPK may reveal no differences between A and B Phases, but the microbiological effects on soils, root growth and plant growth in general can be quite different. A firmer and larger root system with a resulting increase of proteins and vitamins in plants has been observed when Phase B compost is used.

The water penetration is quite different. In other words, the beneficial value of the Phase B compost will show up at the consumer level rather than on the fermentation level.

We emphasize that Phase A compost can be compared with fresh manure, while Phase B compost is very much like the well rotted, old manure. Accordingly, the effect on soils is quite different in each case. It is also obvious that for screening, bagging and storage, as well as enrichment, the Phase B compost will be much easier to handle. It will keep better and remain stable. Legume crops will react much more favorably to it. No nitrogen is tied down in soil.

The B.D. Compost Starter contains, besides bacteria, other important substances such as enzymes and growth hormones which act as stimulant for the bacteria, and as far as enzymes are concerned, also accelerate the breakdown of digestion of the source materials. In this regard, the B.D. Starter is quite unique. It is for reasons of simplification that "bacteria" alone are mentioned in order to avoid lengthy discussions and explanations. The importance of growth hormones or factors and enzymes should not be overlooked. In fact, without these the bacteria could not do their work. All of these factors are an integral part of proper compost manufacturing.

Finally, there is the possibility that sometimes a bacterial inoculation fails. This can be due to toxic ingredients in the raw materials, for instance, residue from chemical treatments, economic poisons, etc. Leather scraps with chromium residue attached will not rot unless the chromium is removed. Another cause may be that the initial heat of the fermentation pile was too high and dried up the pile so that the development of bacteria was interrupted.

Frequent turning and shaking up of a pile interrupt the development of certain actinomycetes which want to work "in peace". If a pile is below the optimum size, smaller than 5 cubic yards, it cannot develop what we call "sweating". Anaerobic conditions and too high a moisture content also cut down the aerobic fermentation. In such cases a pile will lose the most important organisms and only a few not at all essential bacteria will survive. It is for these reasons that we have gone to such length to describe the proper conditions for fermentation.

Of course, in all these instances a re-inoculation with the Starter and restoration of more favorable conditions can rescue the situation, but this adds to the cost and should be avoided by proper handling at the beginning.

CHAPTER VIII

The Addition of Nitrogen Fertilizers to Compost

The suggestion has frequently been made of adding nitrogen fertilizers to composts to enrich the formula as well as to make the fertilizers more "palatable" as a commercial product, in order to appeal to the farmers who have been educated to think in terms of high nitrogen formulas. By "high" nitrogen we understand more than 4, 8, 10, 12, 16% or more.

We have suggested, in the past, enriching a compost formula with a 1 to 2% N content by the addition of nitrogen compounds to raise the nitrogen to 3 and 4%.

According to publications which can be found here and there in the agricultural literature, experiment stations, in Alabama for instance, have found that 4% nitrogen in an organic fertilizer is about the upper limit of effectiveness and no higher yields have been obtained with higher nitrogen formulas when organic matter containing fertilizers (composts, manures) was used. This means that higher formulas only lead to excess nitrogen which in turn will be lost or merely be superfluous.

This observation does not touch the problem of how far one should go with inorganic fertilizers with a high content of easily available nitrogen.

Fertilizer additions of inorganic compounds have been investigated with regard to compost enrichment and this is what we know at present.

Ammonium nitrate is not suitable as an admixture to organic composts because of the danger of spontaneous and internal combustion especially if the compost is dusty, powdery and of a low moisture content. Such mixtures draw moisture easily because they are hygroscopic and then begin to warm up, losing ammonia to a considerable extent.

Sulphate of ammonia, urea, calcium nitrate and Chilean nitrate are under consideration. Calcium nitrate is the best material for our purpose. However, the most common commercial product is a calcium-ammonium-nitrate for which the same rules apply as for any ammonium containing compound, the easy and too quick release of ammonia.

Calcium nitrate is suggested where there is in any case a deficiency of calcium. Microorganisms in vitro and vivo react favorably and especially nitrogen fixing organisms such as azotobacter, nitrosomonas and rhizobium require calcium.

However, the fact should always be born in mind that nitrogen fixing organisms and nitrate formers work best when the available nitrogen is low. Every bacteriologist knows that these organisms are best grown on nitrogen free culture media. If available nitrogen is high they will not fix atmospheric nitrogen but begin to live on the nitrogen which is so easily offered to them, thus even creating nitrogen losses. Therefore, if one wants to make use of N fixing organisms one has to keep the available N low.

In composts where we need these organisms for the breakdown of organic matter and for the formation of a stable humus, the total N has thus to be kept in the lower brackets. It was our experience that a 2% total N and a 1.7% ammonia or nitrate content is the upper limit od N stability. It is possible that the N can go as high as 3 to 4% but then the compost should be dry and not active. During the fermentation in a pile the N can be up to 2% and will remain; with a higher N the danger of ammonia and nitrite formation will lead to losses. Nitrites are especially dangerous because their concentration will lead to a poisoning state and prohibit the growth of microorganisms.

In a low nitrogen compost (between 0.8 and 1.4%), we have observed that with a proper composting technique and the use of our B.D. Compost Starter we have been able to increase the total nitrogen content sometimes by 70%, frequently by 40%. In other words, the actual fermentation is carried out best at lower N levels.

There is also, of course, a lower limit and we have found that compost with a total N level below 0.7% usually is "slow". For the fermentation of very low nitrogen containing materials such as sawdust, bagasse, straw, i.e. cellulose and lignin materials, the addition of nitrogen is mandatory in order to speed up the rotting.

However, the addition has to be done with great care and skill if one does not want to lose too much nitrogen or prohibit bacterial action.

Our present experience is that nitrogen source material with slowly available nitrogen, i.e. organic nitrogen, is far preferable to easily available nitrogen such as ammonia, nitrate, or urea, especially in regard to compost fermentation. In order to add and correct the N situation in a compost pile of low N content as mentioned, we suggest therefore the combination of a small amount of the above mentioned easily available nitrogen sources and a large amount of organic N deriving from hoof, horn, blood, fish, i.e. animal sources or rich plant sources, such as pea vines, castor bean pomace as well as poultry manure.

Occasionally we receive letters from people who report that their compost heaps heated up quickly and high (140 degrees F) for the first 4 to 6 days and then suddenly came to a standstill. This has been observed with the addition of mineral N as well as with poultry manure. In these cases we had too fast a release of ammonia and especially of nitrites which in turn created an inhibiting situation in the pile as far as the microlife was concerned. We can only advise that care be taken with N corrections during the fermentation.

Another matter is the addition of nitrogen fertilizers to the finished product, the dehydrated organic fertilizer which was made from plant wastes, garbage, industrial wastes in general. We have tried the addition of ammonium sulphate and urea. If these mixtures are kept dry they will keep. If they draw moisture, they will heat up and lose ammonia.

Using a mixture of 4% N, half of which derived from added ammonium sulphate and/or urea and applying it to lawns and directly to the root area of tender plants (flowers), has caused burns especially on dry soils, under direct sun, and if watered down thoroughly. Too much N was released all at once. The safest bet would be to use a 3% N organic fertilizer made from compost, garbage, etc.

Most garbage composts are alkaline. Ammonium sulphate would acidify as it does in soil. This may be desirable in alkaline soils and for acid loving plants. In acid soils, such a compost would lose two of its main properties, neutralizing and buffering. Chilean nitrate, containing sodium, brings with it the danger of sodium accumulation, which is very damaging to alkaline soils.

The mechanism, respectively biochemistry, of compost and organic fertilizer application to soils should be fully understood by the farmer and the fertilizer salesman. Mineral fertilizers are concentrates and can be applied in small amounts, down to 100 lbs. per acre, drilled in the furrow, or as high as 800, 1000 or more pounds per acre. Therefore, they lend themselves easily to the use of higher concentrations of 8,12, 16% and more, since the rate per acre can be reduced. The tendency recently is to increase the concentration for this reason. The tendency is also not to mix them with the topsoil any longer, but to apply them to the deeper layers of the soil and, as in the case of straight ammonia, to the irrigation water. Either this, or the ground water in the subsoil will dilute the fertilizer so that harmful concentrations can be avoided. The advice is also frequently given to stay two inches away from the plant roots when applying as a side drill in row cultures.

It is evident that no burns would be produced with high N organic fertilizers if the same methods of application are used. It is, however, and this is a very important point, not in the nature and purpose of an organic fertilizer or compost to be applied as a fertilizer concentrate at 100, 200 or 500 pound level per acre. The composted fertilizer should carry a high percentage of organic matter for all the reasons we need organic matter in soil: as a soil conditioner, to improve the physical structure of the soil, to increase the water holding capacity of the soil, to increase the speed and rate of absorption, to provide food for the microorganisms in the soil, especially organic food which delivers the energy for the growth of the organism and to create specific growth conditions in the Rhizosphere of plants due to the growth of hormones, enzymes, vitamins (B 12), antibiotics contained naturally in a good, living compost.

These intrinsic values in addition to the organic matter and the inorganic nutrients make the compost-organic fertilizer valuable. They cannot be judged solely from the mineral NPK point of view. At least half of the dollars and cents value of a compost is represented by these intrinsic values. Organic compost fertilizers therefore are not mineral concentrates. The more of them applied per acre the better it is for soil improvement. While the mineral concentrates have an upper toxic limit of concentration (lbs per acre) the composts have no upper limit. In both cases there is an economic limit as to cost per acre. The cost per ton is not necessarily a criterium and one cannot make comparisons of formulas and tonnage costs because of the entirely different natures of both types of fertilizers.

The practical application rates of composted fertilizers range between 1 and 3 tons/acre of air dry and dehydrated products and 5 to 7 tons of wet (50% moisture) products or even more, for instance for wet manures a figure of 10 tons/acre is the generally acknowledged application. At these rates the problem looks like this. A compost fertilizer with 1% total N, at the rate of 1 ton/acre provides 20 lbs. of nitrogen, 2 tons/acre provides 40 lbs., and so on. If the N content is 2% these figures double, 40 and 80 pounds of N per acre which is enough in most cases. Of a 3% N compost one would use 1000 lbs/acre and get 30 lbs of N, etc. This explains why it is not necessary from a practical point of view to increase the nitrogen content above the 3% level.

There is still another factor worth our consideration. In composted fertilizers most of the nitrogen is present in form of a stable organic nitrogen. This nitrogen is slowly but steadily released over a much longer period of time then the readily available ammonium or nitrate. Ammonia and nitrate are easily lost in the ground water, in rain or irrigation, as the plant roots do not make use of all of it at once. Only part of the ammonia and nitrate is preserved in the soil, namely that fraction which is absorbed by the natural soil humus or transformed by the microlife in soil. Only when a soil is dry no action whatsoever will occur. Again a soil with a high organic matter will stay moist longer into a drought than a mineralized soil. Many observations in this direction have been made recently. One frequent occurrence is that organic matter (and nitrogen) influence the plant growth favorably still at times nearer the harvest when other plants have already stopped growing, while excessive available ammonia and nitrates would push a plant to shoot up, to produce a lot of green mass, but to lag behind at maturing time, i.e. to ripen prematurely. The danger of lodging of grain exists when there is lots of rain and too easily available nitrogen, a danger which never occurs with the organic treatment.

The farmer and gardener has to become familiar with these fundamental differences, then he will be able to apply the "low grade" organic formula to advantage. Organic N will be much longer lasting in fact in soils with a high organic matter content. The after effect will still be evident in second, sometimes even in the third year, so that a new application is not needed every year. If combined with a conserving crop rotation with legumes this lasting effect can be extended even further.

Due to the entirely different nature and behavior of compost fertilizer in soil the same results can be obtained with low grade 1, 2 and 3% N as is the case with high grade 6,8,12% N mineral concentrates.

CHAPTER IX

Seaweed

Years ago the writer was visiting friends in Wales. When we walked down to the southern shore line, huge carts loaded with a brownish, greenish, bubbling mass came along the road. They were loaded with seaweed which the farmers had collected on the shore. It appeared that this seaweed was very highly valued as compost and in fact these farmers did mix it with other composting material and they built up compost heaps. Also the visit to an inlet showed a peculiar phenomenon. Wild beet roots and sugar beets growing there along the shore line, close to the tide water were particularly lush in growth. A chemical analysis demonstrated that the mineral content of these beets was in proportion almost the same as of sea water, except for sodium chloride. We also learned that these beets will grow particularly well if transplanted into the garden behind the dike, if fertilized with seaweed.

Seaweed has been used in England, Scotland and Wales for many years and was a well-valued fertilizer. There are different types of seaweed. The most important ones are laminaria, also called drift weed or kelp. This plant has a stem and a broad flat lamina or leaf. It grows immediately below low water mark. Its stems are higher in moisture but also they have a very high percentage of potash, while the leaves are somewhat drier and have a lower percentage of potash. The dry stems have about 10 to 12% potash, the dry leaves or fronds, 5%. Another seaweed is fucus; this is also known as bladder wrack or cutweed and grows between tide marks. Its potash content is low, when dried it does not contain more than 2 to 4%. But fucus grows very well in sheltered waters, inlets, where not so many laminaria can be found. Then too, it can be cut from the rock at low tide. Laminaria on the other hand must be washed ashore. Another is ulva, also

59

called sea lettuce. It is washed ashore in great quantities in quiet bays and inlets. This weed is very rich in nitrates. The content of ulva in bays and inlets, where sweet water, rivers, carry a lot of mud, is much higher in nitrogen than one grown in pure sea water.

Seaweed in general, that is a mixture of all of those which can easily be found on the shore, contains on the average per ton 7 lbs. of nitrogen compounds, 2 lbs. of phosphoric acids, 22 lbs. of potash, 36 lbs. of sodium chloride, and about 400 lbs. of organic matter. Barnyard manure, for comparison, may contain per ton 11 lbs. of nitrogen, 6 lbs. of phosphoric acid, 15 lbs. of potash and 380 lbs. of organic matter. Of course, there is no sodium chloride in manure. Fresh seaweed, therefore, is rather similar in its organic matter content as compared with ordinary farmyard manure. It is, however, poor in nitrogen and poor in phosphates, but much richer in potash. Wherever it can be collected easily, with not too much labor and expense, it is one of the most ideal materials for fertilizing and composting. In the old world it has been used before and after potatoes and, in particular, for broccoli, lettuce, peas, and cabbage, also for root crops, mangolds. For cabbage and root crops it is best to have the seaweed well rotted, while for potatoes it could be plowed under in November or December. The ploughing under of fresh seaweed is more easily done when well-rotted. However, if it is to be ploughed under, it should be done immediately after collecting, that is, while it is still wet and green, and early enough so that the salt which adheres to it can be washed out by rain and does not poison the soil. On the island of Jersey, it has been used for early potatoes, and then about 40 tons of freshly collected seaweed were used on sandy soils. It also has been used after the harvest of potatoes for the next crop. When it is dried out about 14 tons per acre were used. The difficulty with drying seaweed is that it deteriorates, and if it is rained on, then many of the minerals, particularly the easily soluble potassium salts, are washed out and lost. If dried, it should therefore be dried on a platform or a floor through which the moisture can not run away, and should also be protected from rain. It is better to dry it relatively quickly, and in the dried state it can be spread directly; but we would advise adding it to the compost heap.

It could, however, be composted in the green state. The observation has been made that if composted with manure, particularly with manure which is very rich in litter, it makes an ideal mixture and aids a speedy decay of the straw. This is carried out in such a way that very little nitrogen is lost and all the other substances are preserved. Its decay is very rapid. The washing out of potash in the case of composting is less of a danger. It is not quite as well balanced as manure, but we believe that in cases of compost, where there is a shortage of potassium, as we have frequently observed in analyzing compost samples, it would be a very good substance to use for balancing compost. For this reason seaweed is excellent for potatoes which need a lot of potassium. Its content of salt might be a certain disadvantage, but composting it or applying it late in fall or in winter, might compensate for the salt. Barley responds particularly well to it, besides the other crops mentioned already, especially on light, sandy soils.

It should not be allowed to rot in large heaps without protection, or without proper composting methods. The other way of preserving it is drying as described. A third way is to burn it and use the ashes, as is done with kelp. The ashes, of course, are much richer in minerals in percentage than the original seaweed, because through the burning and eliminating of water and organic matter, there is a concentration of salts. The potash content in such a case might run up to 12%. In burning seaweed one should take care that it is not heated up too violently, otherwise more volatile substances and trace elements which are present in seaweed might be lost. The potash content of the ashes of laminaria for instance is 28%, while of fucus it ranges between 0.2 to 0.38%, as a dry seaweed between 1.1 and 1.5%. The organic matter of the fresh seaweed ranges between 12 to 25%, of the dry seaweed from 64 to 79%. Its particular value for the root crops and sugar beets has suggested that in cases where the seashore is not near enough to collect the material, one should try to get the dry seaweed, have it pressed in bales like hay or use the ashes as an addition to compost and manure, which is used for root crops. Either in bales or the ashes could be shipped relatively easily. We do not know, of course, how the expense account would be in such matters, because it has not been investigated; but if any of our Organic Gardening readers is interested in the subject and could investigate it, I think he would be doing a valuable service.

Seaweed is also rich in iodine; in fact the burning of seaweed in Normandy and Brittany and in England was one of the important sources of iodine. On a present visit to Rhode Island we picked up several varieties of seaweed and analyzed them in our laboratory. The figures found there confirm the figures quoted about this in the literature. In composting it will be best to interlayer it with earth and other composting material, especially if wet fresh seaweed is used. Dried seaweed would be composted just as one does hay or grass cuttings. The ashes would be sprinkled into the compost heap like thin layers of lime, as they are usually applied. Besides sugar beets, asparagus is a plant which responds very well to seaweed. There the fresh seaweed would be used as well as the ashes, and the sodium chloride content would be even beneficial, for asparagus is a plant which grows much better when a teaspoon of salt is given to each plant. Dry seaweed could also be used as a mulch around asparagus. While its use might not be in question for the inland farmer, it is certainly a very welcome supplement for the farmer and gardener near the seashore.

CHAPTER X

The Biological Role of the Earthworm

Humus is the basis of the fertility of the soil. But not all substances called humus are "humus". There is only one kind which provides the most valuable substance and a desirable soil structure. The scientist calls it the neutral colloidal humus, neutral with regard to its reaction as compared with an acid or alkaline state. All cultivated plants thrive best in a neutral soil, while with increasing acidity more and more cultivated plants cease to grow. They dislike alkaline soil just as much.

The term colloidal signifies a state of matter in between solid and liquid which exists in all living organisms. Only in this state can the "living matter" be maintained. The presence of neutral colloidal humus therefore is a necessity in order to maintain the soil life as well as the proper plant growth. In a soil rich in this kind of humus, plant roots develop the fine hair roots which contribute so much to the intake of the nourishment. The humus is produced by a fermentation process which takes place in a living soil. It is composed of decaying plant matter, leaves, dead roots, stems, straw, and also of manure, compost and of the bodies of microbes, bacteria and algae living in the soil. One animal is devoted exclusively to the humus production —the earthworm. Charles Darwin, in his book about the earthworm, said: "Without the earthworm, no tillable soil." This little innocent and rather helpless-looking animal is our best soil friend. Due to its activity we have a tillable, rich soil. Otherwise, the earth would be covered with rocks, stones, silt and all kinds of diluvial and alluvial deposits, but very little higher plant life could develop. The study of the earthworm is fascinating because of its value to farming and gardening. The presence of the earthworm and the quantity thereof indicate the quality and state of cultivation of a given soil. Its absence indicates serious trouble, an "unhealthy" state of life, less productivity or even a tendency towards a sterile and unfertile soil.

It is clear that the earthworm does not belong in a hotbed, a cold frame or a flower pot, where it may do occasional damage by uprooting or drawing down small plants or seedlings, but otherwise no harm comes from the earthworm's activities. Should it happen that in a garden earthworms attack live roots or seedlings, then something is utterly wrong with the state of affairs in this garden. A deficiency in humus, a one-sided fertilizing with excessive minerals or .lime, fresh manure in bad shape (putrefication) may disturb the proper living conditions for this little friend of ours.

The earthworm is a specialist, and as such needs proper attention. Of the different varieties and species, two groups are of particular interest to the gardener. Type number one is a long bluish, rather thin type, which transforms all organic matter in the soil into humus, and particularly so if the soil has already undergone certain changes. Type number two is rather short, thick, reddish and is the manure and compost worm. Its main duty is the transformation of fresh manure and compost and other matter that is just in the first state of rotting. Type number one would not touch, for instance, fresh manure, but makes the best possible use of decomposed manure and compost.

How does it work out, this transformation of organic matter into neutral colloidal humus? The worm eats earth and small parts of leaves, straw and all kinds of organic particles in the soil. Both earth and organic matter are mixed in the mouth and intestines and digested with the help of an intestinal juice rich in hormones and digestive ferments. One peculiarity speeds up the process, namely, "saliva" glandular juice with an organic calcium compound. The final result of the rather complicated digestive system, full of wisdom, is the neutral colloidal humus. The earthworm excrements, the so-called castings, are the richest and purest humus matter in the world. Analyses have shown that they contain more than ten times as much of valuable plant nutriments such as nitrogen, phosphates, calcium, etc., as the surrounding soil. A Polish scientist, Niklewski, has devoted much study to the humus contained in the earthworm castings and has stated their richness in humus colloids and their importance for the development of fine hair roots.

As to the efficiency of humus production, we look upon it as a sample of the highest possible grade of perfection. In a soil in good healthy condition, for instance, a pasture rich in clover, or an alfalfa or clover field at its best, as many as 300 pounds of earthworms to the acre have been counted, producing about 30,000 to 40,000 pounds of humus an acre per year. No other means of fertilizing or manuring would ever be able to maintain and improve the fertility of a soil to the same degree as the earthworm.

If we figure out that one acre of topsoil to a depth of seven inches weighs about three million pounds, a proper earthworm population digests about one per cent of this soil per year. Besides the humus production in castings, there is the forming of little channels, which increase the porosity and aeration of the soil. This helps other kinds of soil life, stimulating, for instance, about 600 pounds of bacteria per acre a year to contribute their share. Experiments have shown that through earthworm activity the volume of a soil is increased by one-third, which gives an idea as to the loosening up or crumbling of the soil.

Furthermore, the little fellow penetrates downward to about five or six feet, transporting lime, minerals and salts gradually to the upper layers. Many substances previously washed out are in this way brought back to the topsoil. This may explain why soils rich in worms have a considerable increase in calcium content.

Under these circumstances, nobody will doubt that the proper use of earthworms is one of the most important questions of any biological farming or gardening method. But how can we preserve or increase these helpers of the soil?

Unfortunately the life cycle of the earthworm is rather limited. The little fellow is highly specialized. A type number one worm will not propagate in a fresh manure heap, neither will it migrate and accustom itself to a clay soil from a sandy soil or vice versa. It is locally soilbound. The type number two, the manure earthworm, will not multiply in soil except if sufficient manure or compost is present. It is in our interest, therefore, to assist the worm's life. Proper tillage and aeration of the soil, together with the right moisture conditions, facilitate the reproduction. If well-rotted manure or compost is added to the soil, you will see the little fellows multiply as long as they find moisture and warmth. In the fall before frost and during an extended drought they migrate to deeper quarters and remain inert. But before that time, if well provided with rotten manure and compost, they will propagate and at least the egg capsules remain for future activity. The crop rotation also influences the development of the earthworm. Old pastures and fields with leguminous plants like clover, alfalfa, beans, peas, soybeans or lupins are usually rich in earthworm life; while a crop rotation with few of the abovementioned plants and with successive crops of grain gradually depletes the humus and therefore also the soil life. Too great an acidity or alkalinity has a similar effect.

The manure-compost earthworm develops best in the manure or compost if the heap is in direct touch with the soil and covered with earth, and in the case of compost heaps if earth is interlayered with the compost. As long as proper moisture and warmth conditions are maintained, this type of worm will propagate and gradually transform all manure and compost material into humus. The limits of the worm's lives depend upon drought, low and high temperatures. Heaps which heat up too much, above 125 degrees F., are destructive to the worm, as well as frozen heaps and soil. While the southern climate in general is less favorable to the "earth" type of worm because of its periods of drought and heat, it is very helpful to the compost type as long as there is enough moisture, and this can be maintained by watering the manure or compost heap.

The earthworm population of a compost or manure pile can be increased by adding earthworm egg capsules. The late Doctor G.S. Oliver of Los Angeles, California, and his successors, have developed a process which is very successful for breeding worms. If proper conditions in a heap are maintained the worms will propagate by the thousands. The capsules are produced by a peculiar process. The worm hatches the eggs first in the mouth, then produces by means of perspiration a kind of membrane around the head part of the body like a hose. Into this hose about six to twelve eggs are placed by gradually backing out of it and sealing it up. While worms cannot be transplanted from one kind of soil or compost into another, it is very easy to ship the egg capsules and insert them wherever they are needed. Areas in which there is an abundance of roots and of certain plants and trees are preferable as breeding grounds of both types of earthworms. Particularly oak, birch and locust trees, elder and alder shrubs belong to this group. Of annual and biennial plants the already mentioned leguminous family, as well as chicory, onions and valeriana, attract the worms. The addition of leaves or root earth from these plants to a compost heap will greatly reinforce the biological activity. The author has made interesting experiments in order to study the growth and living conditions of worms which squeezed out from the experimental boxes and disappeared. When valerian blossom juice in a high dilution was sprayed onto the surface of the earth no more trouble was encountered due to the worms escaping. As a matter of fact, the worms not only propagated more abundantly, but they also migrated towards those areas sprayed with valerian extract. The Bio-Dynamic treatment of compost and manure reckons with all these peculiarities of the earthworm and stimulates its growth and reproduction.

It remains to say a few words about intensive application of insect sprays containing lead, arsenic or copper. These are biologically poisonous and counteract the development of earthworms. In vineyards or orchards treated intensively for many years with these sprays no earthworms can be found.

CHAPTER XI

The Earthworm and the Toad

About 110 years ago, English gardeners who brought their product to the market places would find there salesmen who raised and sold toads. There was quite a lively trade in toads going on at that time, not only in England but also on the Continent, especially in Belgium. It seems that toads were "planted" in truck gardens because it had been observed that the toad trodden soil produced better and healthier crops.

Another "superstition" would be the opinion of our modern agriculturists. That this was not a superstition may be proved by the importation of one of the largest toads, Bufo Marinus, or Jamaica or American Tropical Toad to Hawaii in 1932. C.E. Pemberton reports in *The Hawaiian Planter's Record*, Volume 38, Numbers 1 and 3, 1934, a thorough scientific publication of the Hawaiian sugar cane planters, how he brought toads into Hawaii in order to combat insect pests. The toads were famous for their tremendous appetite devouring insects in the sugar cane cultures in Puerto Rico. 154 toads were collected of which 149 arrived in good condition in Hawaii. These have since multiplied and have become valuable hunters of insects. They were 13 to 19 days en route with no food at all and must have welcomed the new home and hunting grounds. At first, for a few years they remained inconspicuous and then appeared in great numbers, increasing rapidly thereafter. A substantial decrease of damage by changa (mole cricket) and white grubs has been observed as a result of this involuntary transplantation. "This toad will eat almost any insect that comes within its vision, which includes beetles, moths, wasps, flies, caterpillars, mosquitoes, etc." reports C. E. Pemberton.

In 1716 Peter the Great bought a collection of toads for 15,000 guilders. The Bufo marinus is a giant toad, but there is no doubt that also our native, smaller, northern variety, Bufo vulgaris, as used by the English gardeners, is a very useful helper. Toad hatcheries have been started in Hawaii and it has been estimated that the animal eats about 1000 to 2000 insects until full grown. "We have fed them grasshoppers, armyworms, beetles, earthworms, scorpions, wasps, moths, bees, house lizards or geckos, ants, many sorts of caterpillars, snails in the shell, cockroaches, and sow bugs or slaters; all of which were taken with little discrimination. We read further, "A female bee of the same species was later given to the same toad and was also immediately snapped up and passed down the insatiable Bufo gullet with ease. This large, black carpenter bee, so commonly found boring in redwood fence posts and other wood structures in Hawaii, possesses an exceedingly powerful sting which is used effectively at the slightest provocation. Undoubtedly the bee will continue stinging, for a few moments at least, after passing suddenly into the interior of a toad. It is astonishing that a toad will without hesitation, gulp down a bee possessing such a formidable weapon of defense and apparently suffer no discomfort. After swallowing such a fiery creature as a carpenter bee, Bufo was observed to execute a few abdominal motions suggestive of the Hawaiian "hula dance". Spiders were always eaten with relish if they came within reach of the lightening-like flash of the toad's tongue. Dr. Williams found that freshly decapitated centipedes, which are still able to crawl, are eaten apparently without fear of injury; but the many legs of centipedes are usually able to cling over the mouth or head of the toad so tightly that the process of swallowing such a long object is often accomplished only after much laborious gulping and pushing and pawing with the front feet in a very human but 'inelegant' manner."

One important feature about the Jamaica toad, however, escaped the attention of the research men in Hawaii, and that is the fact that the toad has glands in the skin which excrete a tremendous amount of adrenalin. In fact, the Jamaica toad contains four times as much adrenalin as an adult, full-grown human being. We know that adrenalin is an important stimulating factor in human life. Adrenalin also stimulates plant growth. So we have another factor, aside from the fact that the toad devours insects, namely that wherever it touches plants or seeds or even soil, it adds adrenalin and probably other

69

growth hormones to the soil and plants. Unfortunately this feature has not been investigated as yet. But here it links up with the old English experience. There the toads were not bought and used so much as an insect control, but they were bought because their growth stimulating factor had been observed, and it was seen that crops did much better when toads were around. I have come across this characteristic years ago when I was first invited to visit the farm of a biodynamic farmer in Maryland. This gentleman had quite a row of compost heaps. Some of them were made in a very untidy way, old fashionedly, that is, in comparing them with the properly set up compost heaps, in layers as we described recently. I was asked to find out which of the heaps were biodynamically treated and which not. On the surface, all the heaps looked alike because they were all covered with a thin layer of earth. Now in our composting procedure we prescribe a kind of pit under the heap for aeration and the collection of water. And going around these heaps, I observed that in the case of some of the heaps toads were sitting in those holes, in others not. Remembering the English custom, and also some other observations which will be reported later, I guessed that the heaps where the toads were, were the properly treated heaps. And in fact, this was true. Evidently these toads preferred those heaps for their residence. In other words, whenever you find toads in or near a compost heap you can be sure that these compost heaps will be in perfect condition.

Now, stimulated by all these observations of toads, we made certain experiments before the war on the bio-dynamic farm at Loverendale in Holland. The experimental gardener there kept a toad and had a special box where this toad would live, similar to the earthworm breeding boxes. The toad would always return to this box and became so tame that it followed the gardener when she walked around the garden beds. It was used to the name Fritzi and when she called "Fritzi, Fritzi!" then indeed this toad would hop on the path so that it could be seen. The earth which was trodden by this toad, in the box, was used for plants, for seed germinating experiments, as well as for plant growing experiments. And we observed that the particular growth stimulating factor which was excreted by this toad probably in the form of a kind of mucuous, adhering to the skin, and also in the excrement and urine of the toad, gave an increase in growth in germinating quality of seeds, that is, a seed which probably germinated only 75 to 80% would increase to 95% germination or even to 100%. The plants would also grow faster and taller. Unfortunately the records of these experiments have been lost during the war and so I must quote from memory. We made quite a bit of use of these toads, and the earth from their boxes, in order to carry on seed breeding experiments. We also observed a beneficial influence here. In other words, in addition to the beneficial influence of the earthworm on the formation of humus in the soil, we have in the toad an animal which adds growth stimulating factors to the soil.

The toad-trodden earth was diluted with water, suspended and even the dilutions, 1 to 1000, still stimulated plant growth and seed germination. This is where future research has to start, in order to find out more details, and probably one day we will be able to call the substance which is secreted by the toad by its proper scientific name. If one talks with old farmers and gardeners in Europe in the very old-fashioned areas one will always find a willing interest when the talk turns to toads. This is another case where science could amend its opinions, namely that "superstitions" are not always superstitions, but there is some reason back of them.

The toad, as an environmental helpful factor in plant growth, particularly in organic gardening might become an important item for the future. Much still has to be investigated in this field, but we do not doubt that there will be a good reason that later on, besides the earthworm cultures, one will also start toad cultures again, in case there are no toads around. The toads live in a certain area in which they move freely, and if the table is set for them they are rather punctual in coming to the feeding place. They will live also on relatively dry land, in waste lots, and in cultivated gardens, once they feel they are liked there and well-protected. Of course for propagation they need small ponds, or basins, where the tadpoles can develop. But after they have laid their eggs, 10 to 15 thousand per toad a year, the tadpoles have developed and changed into toads, they are rather independent. We had quite some fun with Fritzi, she or he, we never could find out which, was rather a Lone Ranger. It did not like other toads. Now one day we caught another toad and thought Fritzi was so alone and should have a mate. Probably we didn't hit this right because they started to fight and they would stand up on their hind legs and really dish it out with their front paws. It was quite a sight I can tell you. A boxing match between Joe Louis and his recent opponent could not have been much better.

There does not seem to be any age limit to toads. In captivity they have been kept for twelve years and still were full of vigor. They have a long tongue which they shoot out of their mouth when they see an insect; and can "pick up" this insect at quite some distance. It seems that they can pick up an insect even at a distance of two to three inches. They will sit there and watch the insect for a while, a fly for instance or a beetle, settling, and suddenly the tongue shoots out to the insect, glues it to the tongue and takes it back to the mouth. The whole action is so quick that you just see that the insect has disappeared and the toad looks quite satisfied. In fighting with others, besides hand work, they also spit. Even this saliva, as has been proved with our experiments, has a growth stimulating effect on plants, even though it might not be so beneficial for animals, small animals in particular. Some people complain that they get an itch or a kind of eczema from touching toads. If one is sensitive in this respect one might use leather gloves. Toads will not breed well in an enclosed area, for instance in a box. They can be kept there after they are grown or half-grown, and even if set free they will return to such a place as a hiding place. It is necessary to give them a rather large area. In our experiment we had the breeding box in a space about ten by thirty feet, fenced in with a small wire fence of screening. We called this our toad garden. They hibernate in compost heaps, as was seen, and it would be wise, as soon as one discovers toads near the compost heaps, not to open those heaps before the spring season has started. It is superstion, however, if people think that toads are bearers of evil, or have an evil eye. There is no such thing, and it is only the fear of people who out of ignorance persecute the toad. They can become friendly and companionable, not necessarily house pets, but true friends of the gardener. Protect your toads is the good advice we can give to you organic gardeners.

CHAPTER XII

Mastitis, its Cause and Treatment

Mastitis in dairy cattle is increasing in recent years. It is a symptom of degeneration of the udder and causes much trouble and many losses. If we figure that there are even only one or two cases in every barn, we may well assume that ten percent of the yearly total milk production is lost.

Its causes are, roughly speaking, wrong management and carelessness. Cattle which are forced by means of increased protein feeding to produce more than their constitution can afford, are more exposed to attack than those which produce very little milk, such as beef cattle. In general the higher the production of milk, the greater the susceptibility and the more need of care.

Mastitis is an inflammation of the glandular tissue of the udder. The first stage may not yet be infectious, but sooner or later a streptococcic or other infection will result. At first there may be only fibrous (sometimes cystic) changes in the udder. Once the streptic infection has settled, it is difficult to combat.

That mastitis is in connection with feeding practices is unwillingly acknowledged by authorities. All advise reducing the grain ratio and protein percentage in case of an attack. Only if the udder tissue is weakened will the infection take place - precipitated by external causes such as cold, draft, moist litter or accidental injury, which are the main trouble-makers. Absolute cleanliness of the udder and teats is essential in both prevention and treatment.

Symptoms: a sudden drop in milk production (usually, not always) for a few days; hardening of the udder; a lump distinctly perceptible to the touch; flocculation in the milk; finally pus in the milk, with fever. The very first alarm may be given by the cow refusing to clean up her feed. Mastitis may well start in one quarter and remain isolated there for a long time. It may just as well spread over other quarters.

What shall the dairyman do about it? Good dry litter, a warm dry floor with protection against draft will be necessary. One can often observe that the cows standing nearest the door or an open window are more frequently attacked. As soon as trouble starts, reduce the grain feed, but give plenty of hay (timothy) and water. If silage is being fed, reduce it also to an amount which is not stimulating. Too much silage, particularly in the fall when cows return to stable feeding, can cause udder troubles too. In case of fever, a "red hot" inflammation and infection, it may be wise to give three times a day, a tablespoon of sulfanilamid in a pint of water.

There has been prejudice against the milking machine based on observations that it spreads mastitis and causes more udder irritation than hand milking. Such troubles can be avoided if certain rules are observed. It should become a custom in the dairy to wash udder, teats, hands and all parts of the milking machine which come in contact with the milk, before the milking of each cow. Use for this purpose a mild antiseptic, such as diversol, HTH 13 or other solution containing chlorine. As the milking machines pass down the aisle from cow to cow, with them should go a bucket containing this antiseptic solution in which the teat cups of the machine are dipped between each milking, also a second bucket of clean water in which to rinse them, and a third bucket of cleaning solution for washing the udder and teats of each cow. In this way no infection will be carried from one cow to another. Naturally these buckets will need to be filled with fresh liquid occasionally when it becomes soiled, perhaps after every ten or a dozen cows. It is handy to make a little platform on wheels for carrying these buckets. The cleaning rags and towels should be boiled occasionally and frequently renewed.

The order of procedure in the milking of each cow should be first to wash the udder, but this should not be done until one is ready to carry forward immediately the following activities also, since a cow will start letting down her milk as soon as the udder is touched, and then she wants to be milked at once. Next use a strip cup to test each teat. If any suspiciously lumpy or flaky milk appears on the screen of the strip cup, test each teat with a bromthymol blotting paper. These are commercially produced now under the name of "test paper for mastitis" (Moore Brothers, Albany, N.Y. is one place to get them). Milk a suspicious cow last and the suspicious quarter last and do not throw the bad milk in the gutter but discard it. If any teats show a reaction, do not milk them with the machine, although the machine can still be used on three out of four teats if they are normal. The one showing a positive reaction should be milked into a separate bucket especially marked "Bad Milk" or with a black rim painted around the top. It is quite possible to do this while the machine is on the other teats. This regular testing is most necessary and helps to avoid any trouble through spread of infection. It is worth the effort and even a backward dairyman will soon understand and appreciate its value. By observing all these rules it is easy to keep the bacteria count well under 3,000.

If the first three to six strokes of milk from each teat into the strip cup are normal, then put on the milking machine at once. It has become a general practice to milk fast. As soon as the cows become accustomed to this, the milking can be done in two to three minutes, even with twenty pounds or more of milk. A cow naturally lets down the milk quickly if she feels well. When she holds back, something is wrong. She may be hurt or spastic or frightened. It may be that she has had a stimulation to let down her milk which was not followed up for a while. This might have been the cleaning of the udder, even the touching of it or possibly some other nervous or "psychic" irritation. I remember one cow who would let down her milk as soon as the milking machine was near her and she heard the ticking of the pulsator. Once started with a cow, you should get through the whole job quickly. If she is thus educated to let down quickly, the milking machine will not harm the udder as long as there is a free flow of milk. Under no circumstances should the machine be left on a cow for more than five minutes. It is prolonged pulling and sucking which will cause udder trouble, and if a cow has not given the milk in that time, she will not anyway.

The present idea is to get the bulk of the milk in a short time by machine and then strip the rest by hand. When the machine is used properly there will be very little milk left, perhaps one pound. But this one pound is there no matter whether the cow gave 25 pounds or only five with the machine. This last pound must be taken out, that is, she must be stripped thoroughly. Usually this is in the upper layers and ducts of the glands where the suction of the machine does not reach. It has to be gently massaged down with perhaps ten to thirty strokes. With a good milker this is sufficient and can be done after one has set the machine on the next cow. Few cows will resist this short milking process. There may be one in twenty which is unsuited to machine milking. One which shows a continued udder irritation should not be forced but always milked by hand. The education of the cows to the machine is easy if the milker is consistent in habit and never leaves the machine on for more than five minutes. The cows will soon learn to like it. The low vacuum type of machines is preferable. It is said that the use of milking machines shortens the lactation period. With the observation of these rules this is not the case. Our cows have nine to ten months lactation periods. If a shorter period occurs, it has other reasons, mainly low feed or hay quality.

The specific bio-dynamic treatment for mastitis consists in the application of Silver Ointment and Marjoram-Melissa Ointment (as manufactured by the Weleda, Inc.) This treatment is still in an experimental stage but has given very satisfactory results so far. It has a constitutional and prophylactic effect by strengthening the udder tissue rather than by eliminating acute infections. It is necessary to remember, however, that no treatment can make good for mistakes in feeding or carelessness of the dairyman. In severe cases, the ointments are alternated day by day, morning and evening, and rubbed with only a slight touch into the skin of the udder. This is done for ten to twenty days, in chronic cases longer. If an udder is caked and hard, also for a week or so before freshening, it is helpful to use lemon juice also. Use a thin slice of lemon and rub over the inflamed, caked part or near the lump. Do this at night after work. It has a cooling and contracting effect and many a caked udder has been rescued by this means from serious consequences.

I have sometimes mixed the Marjoram-Melissa ointment with lemon juice as a "freshening" ointment, beginning before freshening as soon as the udder hardens. Recently we have obtained good results with Unguentine too. This remedy was very helpful with a cow which was a "hard" milker because of muscle spasm. It was rubbed lightly into the teats about ten minutes before milking and its effect lasted for three days. The cow seemed to like it and is now as easy as can be to milk.

As soon as an acute congestion of the udder is relieved or the lump softened, one can increase the feed again.

The treatment of udder troubles needs much patience, but there is a good chance of cure if one starts early enough. One needs almost to sense the trouble to come. It is hopeless in a neglected case or if mastitis is recurrent. In such a case it is preferable to dispose of the cow rather than to have a germ spreader as a permanent danger to the herd. It is, however, a fact that mastitis can be kept under control and has never played a serious role in wellrun bio-dynamic farms.

CHAPTER XIII

Answers to Specific Questions

My trouble is with cows. Ten days ago two of them ate (according to the vet's findings) blackcherry leaves, slightly wilted on account of persistent heat and drought. The Jersey, due to freshen in August died four days later after throwing her calf, which was a heifer from a purebred bull! The vet gave her one gallon neutral mineral oil to flush the stomach. It did not help much. He also gave her 2 glucose injections and three other kinds of pills. Nothing helped. The Holstein who also got it, was not so badly off, got on her legs again the third day and started grazing, although her hind legs were still somewhat wobbly. Of course, she gave no milk. (Freshened on May 1st). Last Sunday she lay down again and won't get up nor eat since. She drinks now and then a gallon of clear water and when I shove good hay in her mouth, she will chew and swallow it. Is there anything I could and should do so the cow will get up again? She was in very good shape, and is still not scrubby and meager, but very phlegmatic.

Sigurd M. Rascher, New York State

Answer:

Dr. Pfeiffer recommended coffee, coffee, and more coffee. "Make it strong and give it to her as often as she will take it. That is always the first thing to be thought of in digestive troubles of animals—made stronger than we would drink it."

Dear Mr. Richmond:

Miss Speiden gave me your recent letter since the problems in it are particularly interesting. I was puzzled by your remark that the sudden conversion to organic fertilizer had an unfavorable result. You speak of the danger of serious deterioration of your trees. What were the symptoms of this? If we could know whether there was leaf roll, spotted leaf, dropping of blossoms or fruit or what other diseases or pests, or a small crop—it would help to find the actual cause of your trouble.

The general principles of organic treatment of fruit trees—including citrus— may be outlined as follows: They all require a humus soil, well covered with leaf mold, sufficient moisture in the soil to warrant a continuous and steady growth, with not too much easily available nitrogen. This in excess or in too readily soluble form, stimulates much leaf growth, but blossoms and fruit tend to fall and fruit does not keep well. Sudden drought, especially after much moisture, interrupts fruit development, making it small as well as susceptible to diseases and pests.

Now there is one mistake many tree growers make. The reasons for the condition of a crop—whether good or bad—are most likely dating from the previous year, which influences the start of buds and the collection of reserve substance. From these, the year's crop develops. This is frequently overlooked and people think of fruit trees in terms of annual plants. If this year the soil is poor and dry, the buds and reserves for next year will be poor too and if the trees are forced by stimulation during this weak year, they are liable to collapse rather than improve, and it will take two years for them to recover. Naturally when people see trouble ahead they give more fertilizer and get a vigorous reaction which makes them believe the situation is safe now, but they will weaken the constitution of the tree and sooner or later there will be a collapse. It is similar to the case of a human being who has been working too hard and starts drinking strong coffee to keep awake. One may get away with this for a while but eventually some other trouble will result—a weakened heart, perhaps. Several times I have observed farms in a high state of cultivation which collapsed at first when the system was changed. It requires much experience to change properly without risks. On farms we have found that the safest time to make the change is after a legume crop or when sod or a cover crop is turned under. With fruit trees we often advise not to force the trees to bear fruit for a year or two. Thus they will recover and increase in strength—but the grower does not like to skip a year, or have a small crop for obvious reasons. However, if he has an understanding of good driving, he will understand that it is better to drive a car 100,000 miles at average speed than 1,000 miles at too high a speed.

In practice this means to change to the new system slowly, only as fast as conditions and the economic situation permit. If one has 50 acres of citrus, it may be best to make a 10 year program and change 5 acres each year. A smaller crop from 10 acres of the grove is economically bearable. Also the extra labor involved in the conversion is easier to handle. Under better conditions one may change 20 acres at a time, reducing the conversion period to 5 years. This thought should not be strange to the modern mind, since industry has shown that it needs 18 to 24 months to convert to war and may need 2 years to convert to peace.

The soil analysis and fertilizer figures you quote are certainly realities— for an exhausted soil. The point of view developed by our agricultural experts is perfectly correct, but it has been developed by the study of already diseased conditions. Our modern scientific agriculture is a pathology. If we consider truly healthy conditions, then the problem looks quite different. Then the tree, especially in abnormal years, can rely on reserves, etc.

Your proposed formula for a compost mixture seems very good and I believe you can go ahead safely with it. However, your figures for potash may be somewhat too low. Since potash is the main basic mineral in all plant growth, any mixture derived from plants and made into compost, will be rich in potash. Also the plowing under of green manure cover crops enriches the soil in potash. With a complete conversion to a biological system, there is very little danger of potash deficiency. The nitrogen requirements will certainly be taken care of by a legume cover crop such as crotalaria, or by manure; and the phosphorous deficiency by the bone meal you propose to add. An addition of blood meal may also add to the value of the compost in nitrogen and phosphorous. And the lime question is solved by ground limestone up to one seventh of the weight of the whole compost pile, interlayering it with the other materials, but always so that it does not come in touch with the manure.

Now to the most important point. With the very fine humus you have prepared, following your prescription, you top dress your soil, or disc or plow it under. With the coming of the hot and dry summer, what do you think all this beautiful material will do? Under the intense Florida sun it will oxidize almost immediately and be rendered valueless. The lighter and sandier the soil, the quicker this happens. Even ten tons of humus or manure under Florida conditions will not last longer than 4 months. Plowed under in fall, winter or early spring they may have time to do a little good, but then the plants are more or less at rest. February may be the best time, unless there is protection that absorbs the direct sun's rays. This may be a cover crop or several inches of cover with leaves, straw, Spanish moss or some such bulky compost or straight leaf mold. Since it will take a year to collect any such coverage, you might try first the early manuring. Then drill in a quickgrowing cover crop and mow this down as soon as the heat and drought start. This will give some protection to the soil and added humus so that it can decompose slowly, giving the tree roots the benefit. If with every effort to get a thick mulch, you should come to a point after a number of years where the soil seems sour or sticky, then you may plow and aerate it again. But by this time the soil will have accumulated a good humus content which will serve as a reserve. Having seen the Florida hammocks and how very quickly a once black hammock bleaches out white leaving only white sand as soon as it is exposed to the sun, I believe that the protection of soil from the sun's rays is one of the most important means of soil improvement.

From your letter and the formulas, I cannot guess the size of the land required to produce 400 boxes of fruit. How is it then when expressed in acres and trees per acre? How many boxes from the full grown tree? How many years to mature a tree? When does it start bearing fruit and how long does it bear normally? Maybe knowing this we can work out the compost business more clearly.

It often seems to me that people ask too much of their land. Of a horse we cannot ask more than it can do; a truck stops when overloaded, but with crops, people are often foolish—and then they wonder that trees get diseased and cease to bear after a few years, or that they need more fertilizer from year to year in order to maintain the same crop level. This situation is only too well known to our responsible people. Secretary Wickard pointed out last year that the situation resembles a leaky boat fastened at a given point on the shore. At the same time that the boat is sinking, the river is rising. Observed from a distant point the boat seems to remain stationary so that people do not realize that the boat is sinking until it is too late.

From the biological point of view we know that it takes time to get ill and it takes about the same time to recover—if not more. This holds good for soil as well as for man and beast, and the sooner we start the cure the cheaper it will be. If we delay until late, it will be a very costly business. Soil conservation benefits ought to be paid for all these maintenance operations, which should be deductable from the income.

Dear Mr. Richmond:

As a whole the results of the soil tests are satisfactory and encouraging and I believe that the continued use of compost, mulch, cover and green manuring crops will gradually bring your grove into very fine shape. I was surprised about the good humus-forming bacteria type and count, which we followed up to the identification of the bacteria. Considering the Florida climate and exposure to heat this is very good.

Your thesis that minerals are taken out of the ground by means of crops removed (milk, grain, fruit, etc.) is true. However, the amount removed by washing out (erosion, subsoil drainage) is greater. The first task is to hold this by humus-building management.

Some minerals get into circulation by the natural weathering process of original rock and stones. The amount thus set free every year is entirely or almost equivalent to that taken away by crops—when we deal with average crops say of 25 bu. of grain per acre in average soil or perhaps 5,000 lbs. of milk per cow on a self-supporting farm. But high efficient cropping on average land may take away more than can be replaced by normal means. In forced cropping, one would need either mineral fertilizing or production methods involving the deeper layers of the soil,—that is—increased earthworm population, longer rest periods between cultivated crops, or deeper-rooting plants such as some legumes. If all this is still insufficient to offset the forced cropping methods, then there will be indeed a limit above which only mineral addition could correct the balance—but these additions would be very small. Large amounts are usually given because the soil has not enough humus to hold easily soluble fertilizers. In other words, our soil and humus conserving methods are the only sound basis in the event that one should need to work with mineral supplements. Organic methods give the only assurance that inorganic supplements will be efficient and economical. You may be surprised at such a statement from me, but with my knowledge of soils, I believe it to be true. The organic problem has to be solved first before we can see how much, if anything, else is needed. As a matter of fact in my own experience, I have not yet found a case where it was necessary to fall back on inorganic materials once the organic processes were functioning rightly. However, during the change from inorganic to organic methods—that is during the conversion period —one may need in some cases such minerals as phosphate or lime as a remedy to bring about a quicker action.

A particularly interesting illustration of this idea is the functioning of nitrogen. Many of the bacteria which fix nitrogen from the air do not work in the presence of mineral nitrogen—i.e. nitrate fertilizer. They do not develop on the roots of legumes if nitrogen is applied artificially, whereas if the organic process is taken care of, they develop abundantly and provide sufficient nitrogen in the soil. Hence the need of legume intercrops.

For your grove you need 8 to 10 tons of compost per acre every other year. Green and cover crops can be used, also all organic material available anywhere. Frequent turning is not necessary if you poke air holes into the heap: keep moist; cover against evaporation and the sun's rays with bags, straw, moss, vines, weeds, or anything available, even wood shavings or planks. It should be possible to get thus a good result without or with one turning. If too wet, open drainage and poke air holes into the heap for a time. Sawdust is good material, that from hardwood is better than softwood. Phosphate and dolomite rock go in thin but frequent layers. You will get best results with thin layers of any kind. Use always the best soil you can get for those layers. Poor soil does not add the needed bacteria and humus formers. According to our test you have good humus formers in your topsoil. Manure can be added when the compost is half rotted and the manure as fresh as possible, interlayering the two in 3 to 4 inch layers.

Cattle Feeding Along Bio-Dynamic Lines

There is no doubt that pasture feeding, during the growing season, and hay feeding during the winter, is the best way of feeding cattle. There is no substitute for pasture feeding, particularly with a view to the health and growth of cattle. There is, however, quite a difference as to the feeding quality of pasture and hay. Quantity alone does not make up a good feed, but the quality is much more important. Pasture feeding should consist of lush grasses and lots of clover and alfalfa. Hay also should consist of grasses and of a large quantity of clover and alfalfa. The protein content of grasses is very much improved by clovers and the protein provided in this way is much cheaper than any supplementary feeding with grains rich in protein.

The quality of a pasture is determined by its treatment—the maintenance of good growth, proper drainage, and aeration of the soil. If clovers are disappearing, the pasture has to be renewed or improved. If mosses appear, the pasture has to be surface harrowed. Old pastures with deep growing roots, unless the roots have reached the lowest horizon, are usually of a higher feeding quality than newly sown pastures. It is therefore in the interest of the farmer to maintain his pastures in the best condition.

The curing of the hay has a decided influence upon the quality of it. That is, even a clover hay which is very, very rich to begin with, can be made worthless by improper curing, by getting moldy, by exposure to the sun for too long a time, and to rain, before it is brought into the barn. In the barn itself it can still lose its quality. It is assumed that a badly cured hay has only 25% of its original feeding value, or even less. Grass silage is therefore preferred by some, because the loss in proteins particularly can be much reduced. In wet years a grass silo seems to be preferable to trying to cure the grasses for hay. On the other hand, the cows also need some bulky material for their digestive systems, a mixed feeding is thus advisable. With first class pasture and hay the production of the cows will be high. There are even cases known where this production could not be improved by additional grain feed. Grain feeding is a kind of substitute for the inefficiency of farming, in order to make up for losses in pasture and

84

hay. In addition it is a stimulant in order to get a higher milk production than would be possible otherwise. Grain should always be considered as a supplementary feeding, an emergency, to make up for what has not been obtained from the pasture and hay.

We differentiate between maintenance feeding and production feeding. The feeding for maintenance comprises the amount of pasture and hay which is necessary in order to keep an animal in good growth, good production, and in good health. The production feeding aims to add to the milk production which is normal with maintenance feeding, as much as is needed to break even on a farm. The additional production feeding depends, therefore, upon the quality of the basic maintenance feeding. If the basic feeding is first class, then not much if any additional feed is necessary. The poorer the maintenance feed, the more supplementary feed is needed. This principle is simple in practice, but it is difficult to figure out exactly how much additional feed is needed above the maintenance level. Also this amount is somehow influenced and controlled by the financial needs of the farmer so that he may break even and not lose.

The general observation can be made that grain feeding is often greatly increased to the outer limit of the constitutional and health factor of livestock—even above it—in order to bring about the highest possible production. This is not justified in all cases, nor is it always financially profitable. A sound average in this regard has to be worked out. The writer made feeding experiments with Holstein cows in Holland before the war. These cows were fed on excellent pasture and with a very good clover hay. One group was continued on this basic feed, the other group was given grain, 18% protein mixture, up to 12 pounds per cow, per day. There was no noticeable increase of milk production. On the other hand the writer also has made observations of cattle which produced about 10,000 to 12,000 pounds of milk with protein feeding. When the protein feeding was reduced from a formula of 1 pound of grain per 3 to 4 pounds of milk, the production dropped considerably and it was rather difficult to bring it up again afterwards. In this case the basic feed as to quality of the pasture and hay was not good. There is also an individual factor, as not all cows respond equally to the basic feeding. Some are able to use it more efficiently, some less. In breeding and selection one ought to keep track of this particular property, because if there is developed a

85

breeding stock which shows a high response to the basic feeding, this stock will be much cheaper to keep and to bring to high production than another one. The cows which eat a lot and produce little ought to be eliminated as unprofitable.

The following figures are based on actual observation; they may, however, be modified in accordance with climate, soil condition, etc. They are used to illustrate a certain principle. If a cow on medium quality pasture and hay produces 7,000 pounds of milk per lactation, it is possible to bring her up to 9 to 10,000 pounds production with a grain mixture of 18% protein, used according to the formula of 1 pound of grain to 4 pounds of milk. The same cow can be improved by 2000 pounds above the basic production if the pastures are improved from moderate to first class quality. In this case, in order to top the 9000 pound improved production, less grain is needed; or if the same amount of grain is given it will be possible perhaps to bring her up to 12,000 pounds. Now it is evident that such a cow will be healthiest at a level which corresponds to the normal maintenance feeding. That is, if her basic level was 7000 lbs. and she is forced with grain up to 10-12,000 lbs., the milk production system of the cow is overstrained and it is to be expected that certain breeding troubles and other difficulties may appear sooner or later. The symptoms are udder diseases, breeding trouble, sterility. These are the ways in which cows react to forcing. An engineer knows that there is a load limit for every material—a bridge, a machine, etc. There is also a biological load limit for each individual animal. The farmer has to develop insight and consideration, to observe his cows to see how far he can go without causing trouble. People who have fed cows for milk production for show records, advanced records and so on, know very well the troubles they have afterward with these cows. They believe, therefore, that it is better to sell such a high producing cow and let the buyer suffer. Any change of feeding, weather, even new personnel, creates much more trouble in the case of such cows than with those kept at the maintenance level. At maintenance level, the cow is most resistant to disease, weather conditions and so on. At her top level she loses something of her resistance. All these factors have to be taken into consideration, and we approach a concept of what I may call the optimum health and production level, which is not identical with the maximum production level. At the optimum production level one will also have a cow which is reasonably healthy and of good breed quality. In the selection of the breed and the lines one is building up, one should consider all the factors and value the cows accordingly. The top production cow is all right for production records and advertising purposes, but it might not always be the mother of a large herd. Scientists nowadays agree that if too high a specialization is reached in any organism, it is necessary to go back to an average animal and build up a new breed. The careful cattle breeder, therefore, will have two types of animals in his herd. The high production type, of top or near top level, and the breeding type of improved average level, will both be present.

The goal is reached in several steps. The first step is to improve the maintenance level and then to see what else can be done, so as not to force the high production level right away. The cows which show the highest production at maintenance level will be the basis for the future breed. Cows at top level have no means of improving further, they can only decline. These are the cows which are sold at shows and auctions and usually the highest prices are paid for them. Such foolishness exists unfortunately. Last summer, the writer made an interesting observation. There are pastures of all types on his farm— poor, average, improved. When the cows were put out on the improved pasture they increased in milk production and a stage was reached when they didn't even look at their grain when they were brought in for milking and feeding in the barn. They had sufficient in the pasture.

Pasture and hayfield improvements are fundamental for all cattle feeding. Older fields need a subsoiling with a subsoil plow besides surface aeration with a light spike tooth harrow or Scotch (chain) harrow. The reseeding of clovers is most important. A manure or compost top dressing on old pastures and on hayfields after the second year at the rate of 5 to 8 tons per acre, is most essential. This application should consist only of well rotted, biodynamically treated manure or compost. If possible the top dressing should be preceded or followed by the harrowing. The practice of short leys is good in many cases for improved and intensified clover growing.

These are a few formulas for moderate climates (not for southern U.S.A.) for seedings which we have used successfully. All are expressed in lbs. per acre:

HAY

On well-drained soils

Red Clover	6
Alsike Clover	2
Alfalfa	6
Timothy	4
Landino	1
	19

On wet, slightly acid soils

Red Clover	5
Alsike Clover	6
	—
Timothy	6
Red Top	3
	20

On sandy loam (author's mix)

Red clover	4
Alsike clover	4
Redtop	2
Timothy	4
Landino	1
	15

Poor land to start

Sudan grass	8
Brome	10
Sweet Dwarf Clover	2
	—
Landino	1
	21

On dry land after crop, after barley, etc.

Cow Peas	60
Millet	6
Sudan	6
for grazing, hay or green manuring	
	72

Hay 3 to 5 years on good soil

Red Clover	4
Alsike Clover	2
Landino	1
Alfalfa	6
Timothy	6
	19

Hay 1-3 years on wet soil

Red Clover	4
Alsike Clover	2
Landino	1
Timothy	5
Redtop	3
	—
	—
	15

Pasture, permanent

Red Clover	2
Alsike Clover	2
Landino	1
Alfalfa	2
Timothy	2
Redtop	3
Kentucky Blue	4
	16

Pasture, permanent		Wet, medium acid soils	
Red Clover	6	Red Clover	3
Alsike Clover	4	Alsike Clover	6
Landino	2	Landino	3
Timothy	3	Timothy	3
Redtop	2	Redtop	2
Orchard grass	3	Orchard or blue grass	3
	20		20

Irregular, hill fields, extremes of drainage (also for hay)

Red Clover	3
Landino	1
Alfalfa	4
Timothy	5
Redtop	2
	15

The supplementary grain feeding should consist to a large extent of home grown grain, which can be improved by the addition of high protein supplements. Home grown grain is mainly corn, wheat, rye, barley. Corn, rye and barley seed have about 9% protein, wheat 12%. At its best the home grown mixture contains only 12% protein; if soya beans or peas are added it can be raised to 13-14%. Soya beans are an ideal supplement. They cause scours, however, and should not form more than 15 to 20% of the entire grain mixture. These mixtures can be ground on the farm or at a mill. Many farmers sell the soya bean seed and buy soya or other, preferably linseed oil, cake or meal, with a protein content of up to 34% as an admixture. The final protein content of a grain mixture should be 18% with an average mixed hay, 20-22% with poor grass hay containing no legumes, or poorly cured. When feeding alfalfa (lucerne) and clover hay it may be reduced to 14% or even omitted. The grain mixture should also contain a mineral base with lime (oyster shell) and phosphatic lime (rock phosphate or steamed bone meal), as well as iodized salt, sulphur, trace elements. Perhaps up to 50 lbs. of this mixture should be added to a ton of the grain—10 lbs. at least should be of salt. Sea salt, if obtainable, is ideal for it contains almost all minerals in a perfectly balanced proportion. Of the grain mixture, one feeds 1 lb. per 4 lbs. of milk; with very high producers, 1 lb. per 3 lbs. of milk.

Dry cows and growing heifers should get not more than 14 fitting ration, 4 to 8 pounds per day, according to circumstances. It is most profitable to introduce an individual, selective type of feeding. The cows are divided into three groups: up to 20 lbs. per day production, as little grain as possible; 20-40 lbs. per day production, use formula 1:4 of the 18% mixture; above 40 lbs./day production use 1:3 of the 18-20% grain mixture. Check frequently whether the cows make efficient use of the feeding and adjust accordingly. Many dairy farms lose money because they feed too much in proportion to the production. The writer has seen dairy farms where twice as much was fed as the cows could make use of. It is much cheaper to use excellent legume hay and use grain only as a supplement, or the final spur to make the jump. One advantage of the high grain prices of recent years has been that farmers have learned to appreciate this truth.

Below are a few grain formulas which we have tried with good success:

Dairy 17-18% protein (where much corn is available)		Fitting ration (heifers and dry cows)	
Corn	1400	Corn	600
Wheat	800	Wheat	400
Brewer's grain	700	Oats	800
Gluten	800	Oil meal	200
Linseed cake	400	Mineral base	90
soya bean or meal	400		
Mineral base	60		
	2060		2090

Use 1/4" screen, using commercial
supplements
20% where there is no legume hay.

Corn	800 or 400 oats
Wheat bran	600 or 400 corn
Soya or Linseed	600
Steamed bone-meal	20
Salt	15
	2035

16% with good mixed
clover hay

Corn	600
Oats	600
Soya	300
Wheat	500
Mineral base	30
	2030

11-12% with first class alfalfa
hay *(also dry ration)*

Corn	800
Oats	800
Wheat	400
Mineral base	30
	2030

In addition to the home grown grain, a 32% supplement concentrate can also be used. It must be understood though, that all supplementary feeding is not strictly bio-dynamic, being only emergency until such time as the bio-dynamic conversion has become fully effective.

Now a word about bull and animal "fertility" feeding. A good service bull should be kept lean with plenty of exercise. Good hay, mainly timothy and other grasses, no alfalfa, is best. Corn is not good at all for it is fattening. The hay is best to keep him in breeding condition. Ground oats, 36 lbs. per day, are a good and stimulating supplement, particularly during heavy service periods. Cows, which are not overfed with proteins, can also be brought into heat more easily. Sometimes germinated oats are an ideal "remedy". Every farmer knows that when there is udder trouble he has to drop protein feeding at once. It has been our experience that with proper sanitation and feeding, not overdoing grain and protein, a healthy herd can be kept without mastitis and breeding troubles.

Below is a list of grasses and their properties:

Drought Resistant: Brome grass, Tall oat grass, Wheat grasses, Lespedeza, Sweet clover, Alfalfa.

Tolerant to Poor Soil: Orchard grass, Canada blue grass, Tall oat grass, Red fescue, Chewing's fescue, Redtop, Rhode Island bent grass, Sheep fescue, Lespedeza, Sweet clover, Mammoth clover.

Tolerant to Shade: Orchard grass, Rough Stalk Meadow grass, Meadow fescue, Chewing's fescue, White clover (slight tolerance).

Hot Weather Required: Sorghums, Sudan grass, Bermuda grass, Soybeans, Cowpeas, Lespedeza.

Cool Weather Required: Kentucky Blue grass, Canada Blue grass, Timothy, Brome grass, Wheat grass, Field peas, Winter Vetch, Red clover, Alsike clover, Crimson clover.

Rich Soil Required: Kentucky Blue grass, Timothy, Brome grass, Meadow fescue, Creeping Bent grass, Alfalfa, Red clover, Alsike clover, White clover.

Tolerant to Sandy Soil: Brome grass, Italian Rye grass, Tall oat grass, Canada Blue grass, Meadow Foxtail, Reed Canary grass, Redtop, Red fescue, Rhode Island Bent grass, Sheep fescue, Bermuda grass, Alfalfa, Winter Vetch Cowpeas.

Tolerant to Wet Soil: Reed Canary grass, Timothy, Meadow Foxtail, Meadow fescue, Canada Blue Grass, Rough Stalk Meadow grass, Redtop, Creeping Bent grass, Fowl Meadow Bent grass, Alsike clover.

The good germination of seeds lasts as follows: Most pasture grasses up to 4 years; red clover 2 years; white clover 3 to 4 years; alfalfa 4 to 6 years; best selected seed up to 10 and even 15 years.

Hog Feeding

The easiest and most economical way to raise hogs up to 220 pounds after weaning is to put them out on good legume or rape pasture and provide a self feeder and plenty of good clean fresh water. To those not accustomed to the habits of hogs it will seem that the hogs are eating their heads off from a self feeder. However, it has been proven through hundreds of experimental tests that this produces the most economical gain. An acre of good clover or alfalfa should produce sufficient pasture from early Spring to late Fall for about twenty or twenty-five head of feeder hogs or six to eight sows with litters.

For the grain feeds I prefer corn or a mixture of wheat-corn or barley-corn ground and kept before them at all times for quick fattening.

For concentrates, I prefer, in the order listed, tankage, meat scraps, fish meal and linseed meal, or best of all, a combination of two parts animal and one part vegetable protein. A generous supply of skim milk or buttermilk when available is excellent. For Fall pigs, or dry lot feeding, the old standby for protein supplement is the trio mixture consisting of 50 pounds of tankage, meat scraps or fish meal, 25 pounds of linseed meal and 25 pounds of good quality ground alfalfa. This too, should be fed in a self feeder as well as a good mineral mixture.

Starting with good thrifty 40 to 50 pound pigs, they should gain 100 pounds on 350 pounds of corn at the rate of about 1-1/4 pounds per day. On an average here in the East, I would say that a weight of 200 pounds in 200 days would be a good goal, although a careful feeder with good pigs will average close to 200 pounds in six months.

CHAPTER XIV

A Dynamic Concept of the Weather

One of the daily small chores of the modern agriculturist is to watch the weather and the weather reports. Everyone knows the story of the old peasant who was famous in the whole neighborhood for his unfailing weather forecasts. When asked how he did it he replied in a moment of confidence that he studied the weather reports very carefully and then forecast the contrary and in this way established his reputation. Although such anecdotes are current, yet the material facts accumulated by the meteorological station demonstrate untiring industry and hard labor and should it happen that the forecast does not hit it right, it would surely not be the fault of the meteorologist but of the "weather".

Consider the many factors, such as heat, atmospheric humidity, the earth's heat radiation, water surfaces, forests, etc. that can influence the air currents, and you can understand how the saying arose "changeable as the weather". For this reason it is often difficult to perceive how the different states of weather have any connection.

There are always two designations to be found on the weather charts: high pressure areas, where good weather is expected to prevail, as a rule, and depression areas with rain, storm and lowered temperatures. These conditions follow in the wake of rising and falling temperatures, by solid and liquid elements penetrating the elements of air and heat.

Let us consider what occurs in an air space stationary over a definite part of the earth's surface: —as the air warms up it becomes specifically lighter and begins to move upward, ie., it becomes active in movement. Now appears the influence of the solid element, the earth, because the temperature varies according to the nature of the land below. Over a heavy clay soil the air warms up more slowly than over rocks or sand. The same is true for the cooling off process. If the sun's rays are no longer reflected from the earth, then sandy areas cool off rapidly, woods and water surfaces less rapidly. This difference in temperatures causes the air currents to rise at different rates of speed.

The watery element is also set in motion, that is, it evaporates. Water is transformed into steam and according to the temperature it saturates the air in varying degrees. Then it condenses into more or less tiny drops and forms what we perceive as clouds. If the saturation point is exceeded there follows the condensation of water into rain.

These rising air currents can be observed from the earth, for instance, as cumulus clouds which on hot summer days sometimes form great masses towering over moist places such as lakes or rivers, and are blown away by the wind. The seashore is especially favorable for such observations for the most beautiful cloud formations will rise, thanks to an increase of temperature, over sandy shores. They appear like a crenulated wall, grotesquely shaped, are caught by the wind and driven towards the land, ever-changing from one odd form to another. These towering masses rise higher where the ocean is most shallow.

Wind rises where the warm air strata move off towards a colder region and its intensity can really become a measure for the differences in temperatures that are being equalized. Thus the wind is the external sign for the energy of the air in restoring its disturbed equilibrium.

Often when a storm sweeps over an area, it is thought to have come from a great distance, when it may, in fact, have started quite near at hand by reason of a difference in temperature—for instance, between the bottom of a mountain valley and a neighboring hilly ridge or again between the temperature of the coast and that of the interior of the land near the coast. How often do we see layers of clouds chasing over not far above the ground while cirrus clouds remain immovable at great height!

These apparently simple conditions due to rising currents of air are being continually upset, on the one hand by irregular degrees of warmth in the air, causing ever new cloud formations to obscure the sun, and on the other hand by the equally unreliable compensatory winds.

95

This gives us the changeable, unpredictable element of the air, although there are also a number of other factors to be considered, as for instance, the influence of the earth's movement on the direction of the wind. On the northern half of the earth's sphere the winds have the tendency to veer towards the North and East, while in the subtropical zone they are apt to turn towards the equator. Furthermore, there is the influence of the barometric maxima and minima, which bring a certain degree of regularity into this chaos of the most varied air currents.

According to a general law of nature, the air is ever striving for a state of balance. If this balance is disturbed, then a counter force arises to reestablish it. What we know as "weather" is nothing else than the disturbed balance of the air above the earth and the endeavor of forces to re-establish this balance, which can, at times, be a very stormy proceeding.

These contradictory efforts for balance often create the most amazing situations. Thus it is an astonishing fact that the greatest frigid areas of the earth are not at the poles but at the deepest points of the tropical oceans below the equator, and in the stratosphere. Ice is formed on the water at the poles, but deep down in the oceans at the equator we find a temperature below zero (28 degrees F). The cold ocean currents from the poles sink to these depths on their way to the equator and are further cooled through the strong pressure of the overlying water masses, while on the surface of the tropical oceans there exists overheating, of course, by reason of the intense rays of the sun. The same play occurs in the atmosphere. On the surface of the earth we have the hot tropical air; in the stratosphere, about 12 miles above the earth, a temperature of -60 degrees C. In the polar region the cold registers on an average only -40 degrees C.

By far the most essential weather feature is what Goethe designated as the fundamental phenomena. This is what becomes externally evident to us through the action of the barometer. With a high barometric stand a strong atmospheric pressure against the surface of the earth is indicated, while a low stand indicates a less strong pressure. In the first instance the atmosphere is heavier, denser, more earthbound; in the other case lighter, more fleeting, striving away from the earth.

With a high stand of the barometer in the so-called high pressure areas the solid element and the warmth element have entered into a more intimate connection, while with a low stand of the barometer in the low pressure, the liquid element and the air element(*) are more generally connected.

In the low pressure area it is mostly a matter of rising air currents that are throwing off their water content. In the high pressure area the air's capacity for absorbing water is almost unlimited. These high pressure areas project, especially in summer, far up into the stratosphere (**). They are noticeable by reason of a pressure effect, while in the low pressure areas there is a suction effect. The polarity of the centrifugal and centripetal tendency shows in the contrast of "high" and "low". In the high pressure area the air currents flow away from the center in all directions, while in the low pressure area they flow towards the center.

Whoever studies the weather charts for any length of time can follow the smallest details of these movements on the isobars and by the direction of the winds.

The main characteristic of the state of the weather is the tension between high and low pressure. The air surrounds the earth body and participates in the movement of this body, but suffers a slight delay when compared with the earth movement. In view of its inner mobility it tends towards making itself independent and from this results the continual formation of small independent air bodies—or one might say air beings—which, enchained in the earth atmosphere, begin their own existence at different points struggling with each other for mutual compensation, i.e. they gradually dissolve and thus make room for other bodies. These bodies are all motion.

(*) The word "element" is here used in the full sense of its ancient Greek meaning, that is, one of the basic constituents of natural life.

(**) The stratosphere is that area of the atmosphere that begins about 10 miles above the earth and which is completely at rest, strange as it may seem. The only movement one can consider in this space is what one may designate as the rising and falling of the whole atmosphere.

They live themselves out in motion, in condensation and rarefaction, in contraction and expansion.

The spiral formed by the low pressure area is an in-turning one, that of the high pressure area out-turning. Of course, mutual influences deform these spirals, they stretch lengthwise, they spread out, taking on the queerest shapes, until they dissolve into each other and are equalized. They do not rotate on one and the same spot, but like a tongue of air, they float over the land.

These spirals of the high and low pressure areas are not the whole story. The tendency of their movement can be learned by observing the separate wind directions. If one looks at high and low pressure areas in the vertical, as if making seemingly a perpendicular cut, the difference is evident. In the case of low pressure the air flows together towards the center and then upward, is sucked in and pushed off into the stratosphere. It is as if an intense outbreathing of the earth were taking place—forces working toward rarefaction, dematerialization are active here with low pressure.

The reverse is the case in the high pressure area: here the spiral turns outward moving in the direction of the hour hand, while in the low pressure area it is counter-clockwise. In the high pressure area there is a continual flow of fresh air from the stratosphere to the earth, which attracts it, and therefore the air can absorb a greater quantity of water vapor.

Were one to seek the center of the forces for these movements, one would have to go to the interior of the earth for the high pressure area and to the area outside the earth, or to the cosmos, for the low pressure area. As effective forces we have on the one side what emanates from the formative forces, and is also shown in the earth movement. On the other side we have warmth streaming to the earth from the cosmos, that is the effect of the sun representing the whole planetary system. The up and down motion of these air beings arises from their continuous struggle amongst and against each other and their efforts to frustrate one another. (In former times it was fiery dragons and ice giants and other mythological beings who symbolized these processes.) The course of this compensatory process can be studied in the cloud forms by noting the transforming tendency. If cumulus clouds spread over the whole horizon and transform into a cloud strata, with quite low, vapor clouds, called nimbus, drooping towards the outside, then the center of the low pressure area is coming nearer as the winds whirling around the center become ever stronger. The opposite transformation leads to the high pressure area—from nimbus to strata to cumulus. These in turn dissolve into those extraordinarily long strips of clouds, called cirrus.

By combining the direction of the wind with the tendency of the clouds, one can often, without the aid of a barometer, come near to determining whether one is in a high or a low pressure area. Goethe once said it would be an important study to ascertain one's position from the phenomena alone without using a barometer: "Because everything in this infinite All is in an eternally secure relationship, one thing producing the other, or being alternately produced, I sharpened my sight in the sense of what the eyes could take in and accustomed myself to have the facts of atmospheric and earthly phenomena agree with the indications of barometer and thermometer, without having such instruments always at hand".(*)

Goethe ascribes the up and down movements of the barometer to a higher or lower gravity and compares them to the in-and-out-breathing from the center to the periphery.

Although the atmosphere of the earth participates in the movement of the whole earth body, receiving its impulse from it, yet we must consider as an essential phenomenon the independence of the individual air bodies. Thus it is interesting that numerous high pressure areas floating over Europe and a part of the Asiatic continent originate in the Azores. Just as the mouth of a smoker forms a smoke ring which then wanders off independently, so the Azores are a point where high pressure areas take shape and then wander off. One might almost say that this is a zone of intense inbreathing of the earth, for in the high pressure areas, as mentioned, new air masses are continually coming from the stratosphere. For the Western Hemisphere, the Bermudas fill a similar role causing "Highs"; which as a consequence produce long drought periods in summer.

Every section of the earth has a point where there is in-and-

---

(*) Goethe's *Natural Science Series*, issued by Rudolf Steiner, Vol. II.

out-breathing and the weather phenomena there mirror to a certain extent the characteristics of other natural phenomena of this region.

Compare, for instance, Europe and North America. In Europe we find rapid changes in the succession of high and low pressure areas, in condensation and expansion; changes in the landscape and the geological structure at short intervals, besides a number of different peoples, languages, and dialects.

In North America there is a strong tendency towards uniformity, coordination, or to express it in modern terms, towards rationalizing and standardizing; even the landscape is much more uniform over large areas (as for example the great plains). Consequently, the high and low pressure areas have greater extensions, wander more slowly and stay much longer over the same locality. As a contrast, hurricanes arise representing a point in the low pressure area with a suction effect confined to a very small diameter, and which rotates with great rapidity following the dropping temperature.

In this connection the great mind of Goethe gave the thought direction also: "The increased attraction of the earth, of which we learn from the rising barometer, is the power that regulates the atmosphere and sets a limit for the elements."

The atmosphere tends to flow from west to east. Humidity, rain, downpours, waves and billows, all flow eastward in a more or less stormy manner and from whatever place on their path these phenomena may originate, they are born with this tendency to stream eastward.

Here we can point out an important and often a serious matter. If the barometer has been low for a long time, it does not immediately respond to a rising movement, but continues for a while in the way it started and only gradually, after the upper skies have long ago reached a quiescent state, does the turmoil of the lower spaces attain the desired equilibrium.

Goethe teaches in his color theory how to follow up the way of the light and of the darkness; to recognize the colors as they compensate each other against matter. The force that created light and darkness in the world also created what causes contraction and expansion in the air. Although this is on a different level it proceeds from the same primary spiritual laws.

Goethe says further: "Similarly we have now on the one side the force of attraction and its phenomena gravity, against it on the other side the force of warmth and its phenomenon expansion as independent opposites. Between the two we place the atmosphere—a space, so to speak, devoid of so-called solid bodies, and according to the effect of these two forces on the fine materiality of the air, we have what we call "weather". This element in which and on which we live is determined by the most manifold and at the same time most definite laws."

CHAPTER XV

Does Bread Nourish?

Many people may be surprised at the question, "Does bread nourish?" It certainly was nourishing in the times when ancient people rubbed their corn between stones (as primitive tribes still do today), mixed it with water, and dried the resulting flat cakes either on hot stones or in the sun. It may be nourishing even today in the case of southern peoples, where it is the chief article of food. With us it is no longer our main food, although it is a very important one, at least if we think of breakfast and supper and the amount of corn meal cereals and macaroni used in addition to bread. Yet the question for us is not only what percentage of bread we use, but how the bread itself comes into being.

Bread is made from grain, from the seed of it. The whole potential energy of the plant is concentrated in its seeds. They are able to produce a new plant bearing a number of new seeds. All that the plant has accumulated by means of light and warmth (i.e. energy), and also in the way of matter, by assimilating carbonic acid from air and water and salts from the earth, is concentrated in the seed. In the biological sense this seed represents the total life-energy. Its life forms the nourishment that is supplied to our bodies through the digestive process. Wheat consists of the food-stuffs such as salt, protein and starch, plus the energy introduced by the germinating process.

Actually, then, raw, fresh-ripened wheat of 100 percent germinating power should have the highest food value. In Russia, a dish of this kind made of grains of wheat soaked in water and lightly fried in butter, is eaten. Anyone who has tried this knows that only a few spoonfuls of it are enough to satisfy one's hunger. But our digestive organs have become weak, we are told, and they cannot stand such strong original nourishment. It has to undergo more preparation. As civilization has progressed, so has the character of bread altered, and there is not much more nature-force left in it since technical science took a hand.

Efforts to cultivate wheat, for example, in as many places in the world as possible, have led to its being grown under conditions where it cannot ripen properly owing to the climate. This already means that the quality is lowered. For technical reasons (e.g. with a combine-harvester) harvesting and threshing is done in one operation. But the old-time farmer knows from experience that maturing in the ear means a special improvement of the baking quality of the grain. Wheat must mature; this gives at the same time "a final polish". Fine processes of fermentation then take place, after the original growth process comes to a stop. And in this way a grain is produced that will store well. The wheat seed is a living organism and must be treated as such; that means it must "breathe". It is moved, ventilated, shovelled about, and so on, in order that it may retain its full value. In this state it is exposed to special dangers. If it is not looked after with great care, it will be attacked by vermin of all kinds. It is said of the weevil that it destroys a large part of the harvest on ships and in warehouses. The chemist knows a way by which the troublesome and expensive labor of moving the grain can be avoided—by gassing the warehouse. Amongst others, cyanide gas is much liked on account of its poisonous nature. Now the question is: the vermin are dead, but is the grain undamaged? It has been proved that it is often impossible to remove entirely the gas used, because it has been absorbed. There is a possibility of great harm being done in this way.

The chief cause of alteration of the original quality is the grinding. We know that the earliest method of doing this was by pushing a heavy stone to and fro by hand on a flat, oblong stone surface. A coarse, dark meal was the result. The flour and bran and the hard outer cellulose layer as well, were mixed together. The whole of the ground-up grain went into the bread. Such bread is not indigestible, only it must be well chewed. Grinding went on according to this primitive principle—somewhat improved—until steel rollers were used in the large mills. The stone mill was the first improvement. It consists of one fixed stone and another resting on it that turns around. Its diameter may be about 3 to 6 feet. In earlier times, water or wind

103

provided the driving force. This meant that the stones turned slowly, with about 80 to 120 revolutions to the minute. The flour thus produced handles well. The quality of the stones, the pace, and the grooving of the stones are important. Experienced old millers speak of "living" flour. When grasped, it gives an elastic rebound. An important point in this kind of grinding is that the grain is smoothly rubbed to pieces on grain and stone. Very few such mills are in use at the present time. If we visit the windmills in Holland we find that for the most part only feed is ground there and occasionally we may hear "Modern wheat grinds badly; it no longer handles well; it is sticky and messy while being ground, so its fineness and yield suffer." If we look for the cause, we find unsuitable varieties of grain and the exclusive use of chemical fertilizers as well as the use of varieties which have a high yield but a low gluten content, to be at the root of the matter.

Technical science has found new methods; the wheat is ground with rollers of grooved steel at a high rate of revolution. Different kinds of wheat are mixed together. The flour is cleaned and sieved. And now begins one of the most senseless fashions in the domain of nutrition. Fashion generally goes by the looks of a thing and not by inner content and value. And so it is here in the fashion for white flour. I once heard it said in a mill laboratory: "We must continually compare the color of the flour with what we have to compete with. If ours is not lighter, or at any rate equally light in color, we cannot compete." The result of modern milling is the snow-white bread of the large cities, which in many lands also has been introduced to the country districts. The white bread is good to look at, but quite valueless for its real purpose of nourishment. It is just filling, but has no food value; this will be proved in what follows.

We must picture to ourselves the composition of the grain of wheat. It is built up of various layers.

(1) On the outside there is a protective covering, a membrane of cellulose. This is a hard skin with a certain content of silicic acid, and according to experts on nutrition, it is very hard to digest, and it may be dispensed with. It is considered too harsh a stimulus for the intestines and better replaced by the cellulose of the next layer. We should like to mention here that the different methods of manuring the ground have an influence on the thickness and hardness of this

104

layer, and that a proper treatment of the soil may well render the epidermis more digestible. Unfortunately, there has not been nearly enough research into these connections.

(2) The next layer is that of the aleuron cells. Rows of cells shaped like honeycomb are filled with the most valuable foodstuffs: protein, mineral salts, and vitamins, especially vitamin B, which withstands heat. This layer contains what is needed to build up energy in the human organism.

(3) The actual white kernel of flour. This consists essentially of starch that is used to produce warmth calories and to build up fat and flesh in the organism.

(4) The germ. This is rich in aromatic vegetable oils, fat, vitamins, and salts, particularly phosphates. It totals 1-2 percent of the grain, contains 36 percent albumen as against 8-9 percent in the kernel, and 12 percent fat with Lecithin. This last is an important substance for nourishing and building up the nerves and brain; it has 7 percent mineral matter as against 1-1/2 percent in the rest of the grain of wheat and it contains the total dynamic energy for the building up of a new plant.

Now what does the modern mill do? The germ is oily. When stored for a long time, this oil may become rancid, so it is removed at the beginning. The two outer layers, those of the cellulose skin and the aleuron, are not white and so are taken away with the bran. The starch layer then remains. Because of its composition this has a "fuel (caloric) value", but it is entirely LACKING in all the essential SALTS, VITAMINS, and PROTEINS. Instead, it yields a "beautiful" white bread. THIS PROVIDES, HOWEVER, AN EXTREMELY ONE-SIDED FOOD.

On this point let us hear what an eminent specialist, such as Professor Scheunert, the German investigator of vitamins, has to say: The germ contains some Vitamin A. He says: "Because the germ is taken away with the bran during the milling process, the flours generally supplied by the trade, from which the bran has been removed, are not to be regarded as sources of Vitamin A, but must be counted practically free of it". The germ contains a considerable amount of Vitamin B. If it is removed, this portion of that vitamin is lacking in the flour. As Scheunert says: "…. so we can prove in the case of rats starved of Vitamin B that even 0.5 gram of the germ of wheat and rye is enough to start growth, and an addition of 1 gram of the germ produced the best growth. According to this, the germ is among the best sources of Vitamin B that we possess. Vitamin B is often to be found, it is true, yet in sufficient quantity to give high results it is found but seldom." As already mentioned, the same thing occurs with the outer covering of the wheat. Scheunert says: "From this we know that white flour from which the bran has been removed is as good as free from vitamins".

All depends on the grinding, i.e. upon the proportion of bran and coarse skin to fine flour that remains after cleaning. Wholemeal flours are the best. Flours of 75 percent (75 of the original grain is meal) are lessened in value, and "….in the 65 percent ground rye flour and 60 percent wheat flour, no anti-neurotic vitamin at all could be discovered in the test with pigeons."

It is the same with bread as with the flour. Scheunert says: "In wholemeal bread all the vitamins in the grain are retained in the bread. Wheat bread made of 60 percent flour and rye bread made of 65 percent flour, contain no nutriment at all." Generally speaking, the foodstuffs can be made light through the processes of rising and baking, i.e. they expand more. Yet we hear of bakers saying, "the bread has baked dead". In that case it did not rise properly, had not the fragrant smell; and so on, that wholemeal bread has. The cause is generally too strong and too quick a rising, or quick and hot baking. A shapeless and structureless bread results, that crumbles easily; if a piece is squeezed in the hand it shows no elasticity. But first a very serious matter must be dealt with: the chemical improvement of flour.

No snow-white flour is obtainable by natural means. It must be BLEACHED. Trichloride of nitrogen is at present the favorite means for doing this; if new bread made of quite fresh flour is smelled, a good nose can still detect the sweetish-stuffy smell, although this substance is very finely diluted when used. (It may be interesting to know that artificial improvements of flour are forbidden in

Switzerland.) Flours of inferior quality can be "improved" and "prepared" for human use. All these kinds of expedients have one drawback: the baking quality is lowered. Actually there is a difficult problem here. Flour has to rise, either by baking powder or by yeast. Fermenting processes are biological, dependent on a number of external factors such as barometric pressure, temperature, weather, and so on. The proper conditions are not always immediately noticeable and controllable. That means the dough "works" differently. But the customer likes it always to look the same. Stamps and latchkeys always look the same, and bread should do so too! The time bakers may work is limited by law. But the customer wants his new bread every morning, hot if possible, in order to ruin his stomach from the start, even if he does not notice it and puts down later "feelings" to being tired, or to having slept badly or something else! So the baker must bake quickly and evenly. "Away with Biology!" is the solution, and let us go to Chemistry to produce the same every day. So fashion stupidity, convenience, and unsolved questions of quality make a league, with disasterous consequences. Bakers' helps in the form of salts, such as potassium bromate, persulphate of ammonia, and others, are used, and make quick and even rising and baking possible. No account is taken of what happens afterwards.

Let us look at the "remedy".

Bromide salts are used in medical science to quiet patients with nerves. It lowers the consciousness a little. May one perhaps ascribe to bread in our age of nerves a "healing value"? An expert, Dr.Teleky, says he has proved persulphate of ammonia to be harmful and has proposed that its use should be officially forbidden. Indeed in many countries (France, Italy, Belgium, Hungary, etc.) the use of the so called flour improver and bakers' remedy has been forbidden, or limited, on account of the health of the people, or even on technical questions of military nutrition. Holland and Germany, sponsors of mineral fertilizers were chemically minded for a long time. But at last more and more voices were raised   advising caution. Some years ago in Holland, there was a discussion on the increase of bakers' eczema, a disease of the bakers' hand that had become extremely prevalent of late years, and which was traced to the baking "remedies". We read that two doctors said that it is merely a question of susceptibility or allergy. Such people should leave at once the trade of baker. Also it was said that the bakers' remedies were harmless, for the workers who manufacture them did not get ill. The logic seems to me as if one says: "Poison gas is harmless, for the manufacturers do not get ill!" The question arises, how about it if, instead of the external skin of the bakers' hands, the fine mucous membrane of the stomach and bowels were perhaps to show a like "susceptibility"?

In the interest of public health, we can today only demand that these questions should be investigated and tested as thoroughly as possible in an unbiased, unprejudiced way.

That something is wrong with our modern bread we may even gather from all the advertising of reinforced, vitamized flour and bread. The U. S. Department of Agriculture not only permits but even requests the addition of vitamin preparations to the flour in order to restore what has been lost previously. Common sense would like to ask, "Why take it away first and then add it again at considerable expense? Why not learn again how to use the original unspoiled product of nature?"

We read in an American work, "...White bread, which most of us eat most of the time, may not be perfectly safe; in fact, it may be one of the most dubious foods we eat, simply because we eat so much of it day after day and year after year. White bread is suspect on two accounts. The first is, that with most yeast in commercial breadmaking, there are used 'yeast foods'—potassium bromate, and other chemicals.... The French officials consider it harmful and have banned its use in France....

"The second count against white bread is on the score of the poisonous chemicals used for bleaching the flour from which all but a few American white breads are made...."

Professor E. C. Jordan, Chicago University, says: "It is generally admitted that there is no positive evidence that the substances commonly used in flour treatment, such as chlorine, nitrogen trioxide, and benzoyl peroxide, make the flour harmful" ... A committee of the British Ministry of Health...while expressing willingness to recommend the COMPLETE ELIMINATION of the bleaching agents and "improvers", nevertheless state it as the opinion, that chlorine, nitrogen trioxide, and benzo-chloride SHOULD NOT BE EMPLOYED. Mr. M. Labat, a French writer in the Bulletin of Hygiene, says: "The danger of chronic intoxications following the persistent use of bread made with flour that has been bleached and artificially treated by means of chemical improvers is held to be sufficiently well-established to make the ABSOLUTE PROHIBITION of the use of any chemical improvers in France highly desirable."' (A. Kallet and F. J. Schlink, 100 Million Guinea Pigs, 1933).

The demand for bread standardized in appearance, has had consequences besides the chemical questions in the technique of ovens—electric, hot water and steam ovens. It is generally acknowledged that very good-looking wares are achieved, yet the use of old-fashioned ovens with indirect wood firing still produces the most appetizing bread. Everyone may decide for himself whether he tastes with the eyes or with the tongue, and whether his taste and natural instinct have already reached the dead level of the city street, or whether he still has a feeling for natural forces.

We often hear that whole wheat bread is hard to digest. Yet, according to tests by modern nutrition psychologists, this is only so in the case of nervous weaklings, and of people who imagine they are ill. Even in olden, classical times, military authorities knew all about it. They found the efficiency of their troops was reduced when they ate white bread only, as is the case today (in the Great War, in Mesopotamia, and in highway building crews in California and Texas).

We willingly leave everyone to indulge his favorite taste. But we would also willingly help by bringing these facts forward, so that at any rate from time to time, we may look into our nourishment. "Mens sana in corpore sano" was not without reason the watchword of a world-conquering nation. One small fact says a great deal. In two neighboring villages of the Swiss canton Wallis—Ayer and Vissoie—the populations live under similar conditions. But in Ayre only very coarse wholewheat bread is used, such as has been eaten for generations and baked in the old-fashioned oven. In Vissoie, however, new style white bread, baked by the baker is to be had daily. The result has been: In Vissoie, among 36 children ONLY

FOUR HAD QUITE SOUND TEETH and most of the others had 4 to 7 decayed teeth. In Ayer among 800 teeth examined, only 3 had very small holes. There are indeed people living there who still have all their teeth at 60 to 80 years of age. The children looked different too. Those in Ayer looked better and fresher than those in Vissoie. In particular, Dr.W. Kraft reported that in both villages other soft foods were eaten besides bread, but "it is not the presence of soft food itself that is harmful, only the lack of hard food!"

American dentists have found by experiment that one can produce dental decay with white bread.

All these points show the great importance of bread as food for human beings, but they also show the necessity for suitable preparation. That many favorite customs and tastes, such as white color and too fresh bread in the morning, should disappear is painful to many. Yet people do so much for their health, why should they not begin with their daily bread?

Not for nothing was the lovely saying of the Silesian mystic, Angelus Silesius:

"Bread does not nourish us!
What feeds us in the bread
Is God's eternal Light—
Is Life and Spirit too!"

———————

"Das Brot ernaehrt uns nicht,
Was uns im Brote speist
Ist Gottes ewiges Licht
Ist Leben und ist Geist!"

110

Biologic Rye Sprouted

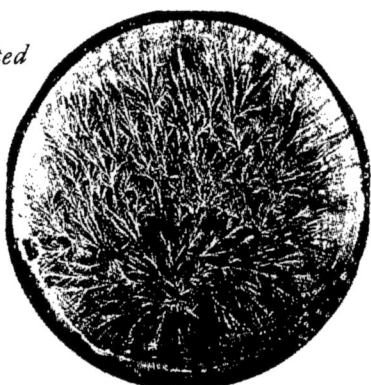

Rye flour with bran

Rye flour

Rye-bread with bran

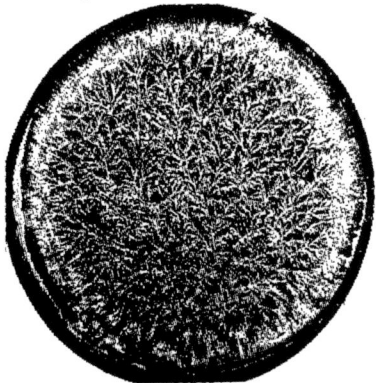

Rye-bread without bran

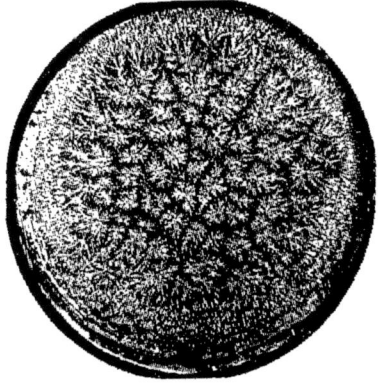

Crystallization of 5% Copper-chloride solution with addition of bran-, flour-, and bread-extracts of bio-dynamically fertilized seed stock.

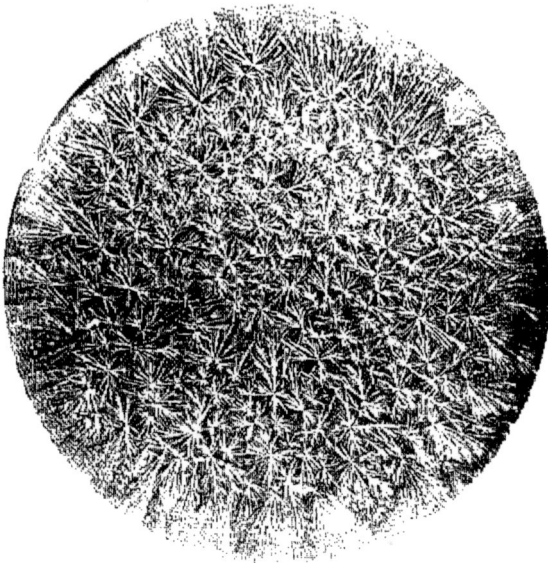

*Copper-chloride with addition of
juice of biologically fertilized wheat.
70% of original*

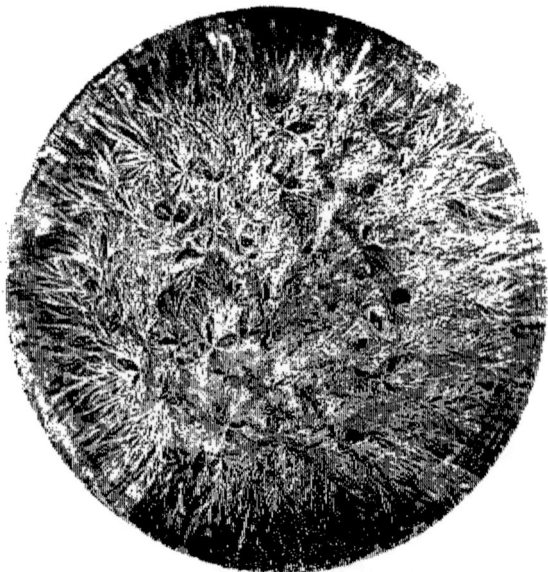

*Copper-chloride with addition of
juice of artificially fertilized wheat.
70% of original*

CHAPTER XVI

Physical and Etheric Energies

Physical energies concern the interaction between mechanical energy, warmth, electricity and magnetism. These are transmutable into one another and are accessible to mathematical reasoning. The laws of physics comprise the interrelations of these energies. The discovery, exact description and application of them have led to the tremendous development of modern technology and industry. The pattern of economic, social and political life and structure has changed more rapidly during the last 150 years than in any previous period of human history.

It is inherent in the nature of these laws and energies that they are valid over the entire earth; they are truly cosmopolitan and are not tied to any nation, folk, tribe or social structure, whether they are used or misused for any purpose. In themselves they are objective.

In so far as a living organism—plant, animal or human body—participates through growth, maintenance or breakdown in the functioning controlled by changes and reactions of matter, these energies in the living realm reveal also the laws of nature.

In addition, there is gravity as another physical energy. Gravity stands by itself and is not convertible directly into warmth, electricity or magnetism; it is static. It can cause movement—free fall or the flow of water, for instance—but movement in one direction only: towards the center of the earth. A direct anti-gravitational field has not yet been discovered in the mechanical realm.

Although gravity shows certain degrees of intensity (the gravitational pull is different at different locations of the earth), there is no means of increasing it or of screening it off. It can be overcome in a fashion by way of thermal or mechanical energy, but it is not changed by this. Gravity is a property of matter.

There is another energy, commonly described in terms of warmth, electrical charge or otherwise—the energy of light. The description of light in the terminology of the other physical energies is obviously incomplete. Although these energies form part of light, they do not make up everything which is conveyed by light. Light is more than the sum total of all other physical energies. What we observe as "light" is only the effect which it brings about when it comes into contact with matter; colour, for instance, arises in this way. Light itself, as a pure entity, is invisible; it is an energy or agency independent of this earth. It is of cosmic origin, and would exist if there were no earth.

As far as a body consists of matter, it is subject to gravity. The role of light is more than the physical appearance of it in or about matter. If we had a world consisting only of inanimate mineral matter, we could have warmth, electricity, magnetism, gravity, but no light, no life, no growth. Light creates life and maintains life. The change of dead matter into living matter is due to the fact that light enters the material realm and is partially transformed into living matter or the living process; this can be seen in the plant cell, where light is absorbed and transformed into chemical and other energy by the interaction of the chlorophyll and related substances in the process commonly called photosynthesis. This living process can be neither described nor explained by merely putting together physical energy reactions and matter, although it is true that these accompany and modify the lifeprocess.

The energies which maintain—in fact, cause—life, we call, following the concept of Rudolf Steiner, "etheric energies". We will try to explain what this term means.

In photosynthesis, under the influence of light—or, more correctly, under the influence of the rhythmical change between light and darkness—physical matter is rearranged and combined into a most complicated system of organic substances which form the living cell. Any chemical reaction either releases or fixes warmth; no chemical formula is complete unless the change of caloric energy is included. In photosynthesis, if we take the concept more widely than is usually done, we have to add to the caloric change that particular

114

energy differential which is brought about by light. (*)

We must therefore recognise that the complex living organic system fixes or releases both warmth energy and light energy. Likewise water is split off or enters the molecule. Without the versatile role of water, no life would be possible. Water is not only a carrier (in solutions), but becomes an integral part of living matter. It is seemingly inert outside, but activated inside the cell. Its hydrogen and oxygen components are reactivated or liberated, making them useful for the synthesis of organic matter. The term "organi" is used here in a twofold sense: (a) strictly according to the chemist's definition; and (b) as part of the living process which leads to organic—i.e. organised— matter.

The newer concept of photosynthesis speaks of a light process: namely, the breakdown of water and the absorption of the hydrogen ion in statu nascendi by a suitable "acceptor", in which enzyme systems play an important role. The second process is the darkness process, comprising the breakdown of carbon dioxide, the carboxylation and the building up of a malic acid under the influence of other enzyme systems or biocatalysts—for example PNtranshydrogenase. This is the starting point of the synthesis of sugars and carbohydrates in general, the building up of amino acids, protein and peptide chains, conversion into fats, etc., etc.—in short, the entire carbon-compound cycle.

Certain mineral elements play an important role here, as do iron, magnesium and phosphates. No light-synthesis would be possible

---

(*) Theoretically, as one writes a chemical-caloric formula, $R_1 + R_2 = R_3 + R_4$, plus or minus x calories, one should write a photosynthesis formula: plus light, and should follow the "light" in the continuation of the process into the biochemistry of the living system—for example, the enzymatic reactions based on enzymes with phosphorus or magnesium in the molecule as bearers of "light" action—while the minus light-reaction—that is the darkness phase of the carbon dioxide-malic acid-sugar-carbohydrate synthesis—is the truly caloric or warmth phase. The continuation of the "light" energy into the system is not as logically and consequently followed up at present as is the caloric change, the intake or release of energy. From the point of view of atomic physics, we need to introduce the fact that the electron has mass, though a negative charge, while the photon has neither mass nor charge. It is the photon energy, however, which is basic to photosynthesis.

115

were it not for the catalytic influence of magnesium and phosphorus, the bearers of "light"—energy in photosynthesis. The warmth-energy is transformed and made available in the sulphur-directed compounds, in the accumulation of sugars and carbohydrates. It is here that the caloric value is of importance. In order to synthesize one molecule of glucose, 686,000 calories are required. The oxidation of glucose in the animal body releases this energy again; fifty to seventy percent of it is used and fixed again in the pursuant chemical processes.

One acre of maize, giving a yield of 100 bushels of seed, will produce 200 lbs. of sugar a day. One quarter of this sugar is oxidised in the course of respiration. In writing out chemical formulae the energy factor is usually omitted. To what extent this energy factor enters in may be illustrated by the following example. On one acre of maize about 10,000 plants will grow in 100 days. The total energy available from sunlight (warmth) is 2,043,000,000 calories. Of this total, 33,000,000 calories are used in photosynthesis; 91,000,000 calories are used in transpiration and transportation of water. In the case of maize, 408,000 gallons of water per acre were consumed during the growing period, using up 45 percent of the total available sun energy. In general, in order to produce one pound of plant matter, a plant consumes about 500 lbs. (wheat), 600 lbs. (maize), 800 lbs. (alfalfa), of water. Respiration releases 8,000,000 calories. At maturity the seed contains one-fourth of the absorbed energy(**).

The light equation has not yet been formulated, but from known facts it can be inferred that the entity which is accepted and transferred by enzymes, and activated by magnesium and phosphorus compounds, supports and maintains the continuous change or metamorphosis of the living matter. It appears that plants assimilate chiefly in the morning and late afternoon hours, and rest at noon and during the middle of the night. They seem to be more photosensitive to the red pole of the spectrum, transforming energies into chemical energy. The morning and late afternoon light includes more red rays, while noon sunlight includes more blue rays.

(**) Figures from "Dependence of Plants on Radiant Energy", Earl S. Johnston.
(Smithsonian Series, Volume XI, Part VI.)

The growth process, however, and especially the elongation of stem and root, takes place during the "dark" phase. My own experiments have shown that the root tip grows more in the early morning hours (with a second maximum in the late afternoon) than at any other time of the day. It is a rhythmical process, like everything else in the plant. In continuous light a vegetative growth, with no flowering has been observed. Blue-violet has more of a retarding effect than the red. Chlorophyll is formed in the red light. It has, however, been shown that certain phases of photosynthesis are affected simply by LIGHT, independently of the wave length, while the growth pattern reveals a relationship to the spectral wave-length. This means that in light there is more than just the caloric energy. Wave-length expresses the physical energy. The question then arises: What is the "surplus" in LIGHT, invisible and not accessible to the concept of physical energies?

Morphology in the maintenance of species as well as in chemical reactions, follows a certain pattern. The pattern is preconceived and directional, already present and active when the first "Anlagen" begin to show up. It coordinates all interactions with the terminal aim of producing just this or that organ, leaf, blossom, sugar, amino-acids, protein and peptide chain, etc., and makes the corresponding enzyme systems appear and disappear. We are dealing with a functional, directional system, resulting in the complete organism. The right reaction at the right place at the proper time: this is growth and health. The same reaction at the wrong place and times: toxins and disease. Cancer, viewed biologically, is a growth process at the wrong time and place. At the right time and place, and in the proper sequence, the growth - forces in the wrong position cause malignancy.

It has been shown that light influences the absorption, diffusion and passing on of minerals in the cells, according to a rather specific pattern and relationship. Tomatoes respond better to nitrogen fertilization in a sunny than in a dull, cold season. The same tomatoes respond better to potash fertilization in a cold season than in sunny weather (E. J. Russell, Rothamsted). Bromine was absorbed and transported by nitella cells four times as fast in light as in darkness (Hoogland, UCLA). Doubling the light intensity increased the bromine absorption by 30%.

117

The absorption and transformation of physical energy is only one part of the entire life-process. The proper arrangement of matter, and the form-pattern according to the predetermined principle inherited in the plant, make up the other part of the life-process. It is this other part which we give the name of etheric, formative force. It is the sum total of all the forces which determine the pattern—the pattern of function and form as well as the pattern of chemical synthesis.

These etheric forces are not of a substantial nature, measurable in terms of calories, electrons, etc. They are invisible, but without them there would be no growth, no form, no organization, no arrangement or rearrangement of living matter. They have also been described by others, rather abstractly, as the determining or organising factor. Matter (substance) acts as the carrier of them, but it does not produce them. The facts point to an independent agency in nature. These etheric forces have nothing in common with the hypothetical "ether" of physics.

* * *

For the philosophical understanding of these processes it is well to realise that a breakdown process is at the starting point of life. This may be the breakdown of water and carbon dioxide in photosynthesis; of sugars and carbohydrates in order to deliver caloric energy to the life-process; the breakdown of more complicated peptide chains in seed germination or in order to produce enzymes. According to its proper position in the living system, matter is differentiated and coordinated. The pattern of form, the model or template —many names have been given to the phenomenon in order to get round the acceptance and acknowledgement of an independent etheric energy—is inherited over generations. Chromosomes and genes are carriers of the fundamental pattern, but not the origin of it. The original pattern is an inherent, immaterial activity, the determining force or energy of which we speak here as "etheric". If this force is related to a definite physical body, one can speak of an etheric body. Just as there are magnetic, electric, gravitational forces and fields, so there is this etheric energy as an independent form of energy, superimposed on matter in the living realm, the sooner will science begin to understand the creative force which acts out of cosmic order and wisdom in order to create and maintain life.

The fact that matter has to be broken down in order to give way to a new development is self-evident. Goethe expressed this principle in the beautiful words: "Nature has invented death in order to have much life." Once the existence of the etheric is recognised, it will be so much easier to integrate the thousands of details which are produced by scientific research, overwhelming in quantity, often incoherent and contradictory, puzzling at times. The horizon of our human concepts needs to be widened in order to understand life.

This etheric, formative force and field are part of natural law. They are accessible to logical reasoning and description; to mathematical—and especially geometric—exploration. In fact, projective geometry provides an excellent means of describing them. As any natural law is an expression of the original creative thought, so is the behavior of the etheric formative force an expression of this same creative thought and intelligence. It is an expression of intelligent action, reason, purpose, determination, metamorphosis, and, above all, of the functional, harmonious relationship between all the entities of nature and the cosmos. It is the primary force of an order higher than that of the physical energies. The latter are executive; the first is creative.

As we have clearly seen, light is a creative force; without it there would be no life. Animals and human beings need the transmuted energy which is fixed in the green plant. The transfer takes place by way of nutrition. To look upon food just as an accumulation of matter, of sugars, starches, proteins, fats, is a meagre concept. There is more in food than just matter. The energies which are transmitted, whether they be caloric energies or pure light or creative energy, are as important and as integral a part of the food as is its substantial value.

Foods for the plant are water and air (carbon dioxide, nitrogen, oxygen, etc.) with the minerals acting as directional or biocatalytical factors. Potash, for example, influences the building up of starches; magnesium, sulphur, and phosphorus influence the building up of proteins. According to Professor Lyle Wynd of Michigan State College, one of the gravest errors of modern plant nutrition has been to designate the mineral components in soil and fertilizers as "plant food". The plant's food is air and water. A new concept in agriculture and nutrition will result from this changed point of view. "Feed the

119

soil (life) and the soil will provide the necessary elements for the plant."(Wynd)

Digestion is the function which breaks down the prepared combination of substance and releases the energies, both physical and etheric. Then the body's own process takes over and rebuilds new combinations to its own pattern. The etheric body of the individual is the organising factor. As light and warmth are released and become apparent in the burning of coal, wood or candle —a release of the sunlight which was absorbed years or thousands of years ago by vegetation—so energies are released in the body according to a fashion inherent in this particular organism.

* * *

One important law of the entire process is the law of the equilibrium or balance of function. What is meant can be deduced from a wider application of the second law of thermodynamics. The Russian physicist Chwolson expressed this more than 50 years ago in the general form: Nature tends to maintain a balance. Wherever there is a disturbance of the balance, a counterforce appears in order to restore the balance. Rudolf Steiner pointed out that this law has universal spiritual importance, as well as physical importance.

Another important law is the so-called law of the minimum. Derived originally from the relationship between the mineral content of the soil and plant growth (Liebig, Mitscherlich), it implies: Growth is determined by the substance present in the minimum degree—i.e., this substance, whatever it may be, has a limiting effect. Translated into the biological concept of function —that is, of the dynamic relationships between all the factors collecting life —the law can be enlarged and will then include not only "substance" but also "energy" or any factor pertaining to the life-process. The general validity of this law is quite obvious; factors related to growth include not only mineral nutrients in the soil, but nutrients of all kinds - humus, trace minerals, water, oxygen, carbon dioxide, air in general, nitrogen—in fact the entire biochemistry of soil life with its enzymes, growth hormones, vitamins, and also all the vital energies, light and warmth and cosmic influences. We must also take into account inherited properties, the entire phylogeny and ontogeny, and last, but not least, the environment, with its supporting or disturbing influence.

Nitrogen in the soil may become deficient, resulting in a plant of weak growth, low protein content, and pale, flabby leaves. But nitrogen is plentiful in the surrounding air. It requires only the particular type of soil organisms which fix atmospheric nitrogen and make it available via ammonia or nitrates. If, on the other hand, humus and aeration are deficient, the proper bacterial activity absent, this complex of factors may be at the minimum level, and not the nitrogen compound itself. Or the determining influence may be the rhythm of light and darkness, or abnormal weather, with lack or excess of warmth. The plant will then not only grow with a lowered protein content, but with inadequately built-up enzymes and growth factors, so that as food it will convey not only its low protein but also a weakened biological activity, reflected in malnutrition. Abnormal conditions of metabolism are the result throughout.

It is well known that carrots may not contain even a trace of carotene (provitamin A), which got its name from the carrot, because of the neglect of factors related to its development in the plant. Such deficient carrots were created by means of an intensified fertilizer application designed to produce big yield (bulk). They did not have the time to develop finer biological processes which make them nourishing. A strain of carrots was subsequently developed with a high carotene content, but its growth was slower and did not yield the large bulk-weight. Farmers did not like to grow them because housewives did not buy them, wanting only the biggest carrots and not knowing about the carotene content. One might say that here the intelligence and insight of the farmer and the housewife were the minimum—i.e., the limiting factor, not the soil or plant.

If the factor of light and cosmic influence is, for any reason, at the minimum, we may still get bulky substance—i.e., big yields but the vital energies carried on into the food will be low. This is the reason for much degeneration nowadays, for susceptibility to disease, low defense reactions of the body, etc. The etheric component is weakened. Increasing nervous disorders, even inability to take decisions, or to make mental adjustments to the faster and faster pace of modern life, may result.

The other law, of the balance between all factors, applies in the same way, so that we can say: The sum total of all factors and functions is constant. We can easily see that bulk production may reduce the energy fraction. A man thus nourished may fail to get vital energies out of his food, in spite of all the protein, starches, sugars, fats he consumes. He eats a lot, but lives in a state of constant malnutrition. This phenomenon is only too well known, even in the countries of plenty, such as the United States or Australia. Malnutrition in the midst of plenty is all too familiar. If food were grown in such a way as to provide all the essential vitamins, trace minerals and other factors and energies, people might not need to buy manufactured supplements. In the processing of food further valuable elements are lost, so that we need enriched bread and many other supplements. Modern life with its haste, mental strain, insecurity, lack of inner balance and lack of spiritual compensation, adds to the strain on human beings. To reckon only with matter and material values does not exactly help in this situation.

The concerted effort of the living organism to maintain a balance between all factors can be seen in every physiological or biochemical function. One further example may illustrate this principle. Urea is a waste product of human metabolism; it is toxic and must be eliminated. However, urea adheres by absorption to the red blood cells, together with the enzyme urease. Urease breaks down urea into ammonia, carbon dioxide and water. But the plasma fluid, where the red cells swim, contains another enzyme which prevents urease from breaking down urea. Then a balance is maintained, failing which a release of toxins would occur. However, in the mucuous membrane of the stomach urea is broken down, and ammonia released, for a specific reason. Otherwise the hydrochloric acid plus the stomach enzymes (pepsin), which are necessary for digestion in the stomach, would attack the stomach linings and begin to digest these also, as they do with the stomach contents. The ammonia transported by the red cells into the membranes of the stomach linings and released by urease, neutralises the acid and protects the stomach linings against self-digestion. Here the breakdown product, urea, acts as a remedy, applied by the wisdom of the body. A human being who by constant worry and mental strain increases the breakdown forces in his digestive system, and is therefore in want of etheric energies, which would rebuild or protect the stomach lining, does not get the benefit of this restoration and may develop ulcers.

It has been estimated that some 10,000 enzymatic reactions (breaking down and building up) take place every second in the living cell. The regulator of the protective, building-up phase of these reactions is the etheric body, which arranges and rearranges matter according to a pattern, reflecting the original cosmic order and harmony. (The breakdown phase is controlled by what Rudolf Steiner calls the astral forces, but to describe this in detail would be outside the scope of this article). There is wisdom and obedience to law in these functions—more, in fact, than human beings can grasp consciously at present. They are controlled by a power of reason superior to man's. Thanks to the cosmic world-order, the arbitrary will of man cannot as yet enter into or control these socalled subconscious or autonomic functions of the nervous system. Understanding of etheric energy will gradually enable man to overcome the destructive effects of his arbitrary willing, feeling and thinking. Theories and hypotheses have to be replaced by an actual imagination of, or insight into, the etheric realm, otherwise man will only play with the unknown and perish.

One first step is the establishing of a conscious balance of emotions, a control of thought which will pass only such thoughts as are in tune with the cosmic order of the etheric. Such a living concept of functions has immense possibilities. Not only will harmonious processes in the body be established, leading to an entirely different definition of health, but an appreciation of the concept of biological order will also influence the inter-relationship between man and man—that is, the social order. Emphasis is laid on the word ORGANISM. In an organism, we have harmonious cooperation and coordination of all factors and functions, a functional inter-relationship. Today we are living in an a-social or even anti-social world. Groups, nations, fall apart because of conflicting group interests, whether economic, political, national, or racial. To the body social it means as much as to a living organism to find the adequate coordination as a result of a functional concept. If the heart were unwilling to cooperate with the lung or liver or eyes, disease would result. Disagreement between the members of an organism leads to malfunction. Finally, self-destruction ensues. This is true both of the bodily organism and of the body social. Nature teaches us the organic concepts of a superimposed form-pattern and of harmonious coaction. Man's task is to learn from nature how an organism works and then to create the body social accordingly, in tune with the cosmic order. Otherwise a materialistic concept of matter will lead only to atom bombardment, to fission and fusion; that is, to ultimate destruction. To recognise that there is a higher order in man, which he can first come to know through changed ways of thinking, practised by free will: only this can create a truly human society. This is **humanity**. The individual, the group, the nation, the race, are all members of this higher order.

Bibliography

Enzymologie. O. Hoffman-Ostenhoff. Vienna, 1954. Springer Verlag.

"The Enzyme as the Fundamental Unit of all Biological Activity". William H. Bond. *Journal of Applied Nutrition*, Volume 9, 1956.

"Bundesanstalt fuer Qualitaetsforschung Landwirtschaftlicher Erzeugnisse, Geisenheim: *Taetigkeitsbericht*, 1954/55 and 1955/56. Prof. Dr. habil. W. Schuphan.

Grenzen der Naturerkenntnis. Rudolf Steiner. Dornach, 1937.

"Der Entstehungsmoment der Naturwissenschaft in der Weltgeschichte und ihre seitherige Entwicklung." Rudolf Steiner. 1937

"Die Naturwissenschaft und die Entwicklung der Menschheit seit dem Altertum." Rudolf Steiner. 1937

Erde und Mensch. Guenther Wachsmuth. Archimedes Verlag, 1952.

Die Entwicklung der Erde. Guenther Wachsmuth. Phil. Anthr. Verlag, Dornach, 1950.

"Die Aetherischen Bildekraefte in Kosmos, Erde und Mensch". Guenther Wachsmuth. *Der Kommende Tag*, Stuttgart, 1924.

Sensitive Crystallization Process. Ehrenfried Pfeiffer, 1936.

Textbook of Biochemistry. West and Todd. Macmillan Company, 1951.

CHAPTER XVII

Money, and Working for the Work

The 19th Century with its profound development of such sciences as physics and chemistry, together with their application in technique, has been called the mechanical age. But 20th-century science has advanced toward an apprehension of living forces and living matter, of biology, of physiological chemistry and of functions rather than statics, so that today we hear about dynamics in relation to nearly everything. This represents not only a progression in thought, that is, in fundamental philosophy; it has to do with the nature of the investigated object itself. In other words, the study of any living system inevitably encourages this point of view.

We can clearly state that thought has advanced from a mechanical to a living point of view whenever happenings in the outside world are considered as functions of a living unit.

Theoretically everybody will agree that this new viewpoint has helped greatly in the solution of many problems pertaining to basic science as well as in applied and practical fields—for instance, in agriculture. To see the farm as a living organism or biological unit has not only opened the mind to an understanding of how to organize soil protection, together with successful farming, but has taught us to see the new bio-dynamic type of farming as the basis for a rural community—indeed, the basis for a new social structure balancing rural and industrial activities.

The fundamental thing about this method of thinking is that it permits one to see an organic relationship, that is, a functional coordination, between the factors, leading finally to the formation of a unit (whether biological, agricultural, social, economic, or what not).

An intensive study reveals that the human being, that highest result of nature's development in all its functions, acts as a model in the training of this kind of thought for use in other realms. Furthermore, the study of plant and animal life in their mutual relationship, and in their relationship to the earth, leads to a better understanding of the "dynamic" relationship.

Up to the present day, in spite of having already demonstrated its value for many practical matters, this point of view has rarely been applied in the study of problems of economics. When adopted, it has exposed the speaker or writer to the charge that he is remote from reality, a dreamer or even a utopian.

But critics of this point of view forget two fundamental facts:

First, that many of the dreams of a hundred and fifty years ago have become realities. (Think of flying across the Atlantic in less than a day, think of talking by wireless from one continent to another, think of the progress in surgery, and so forth.) In other words, yesterday's dreams become today's scientifically grounded thoughts and tomorrow's practice, inasmuch as thoughts are the power motors behind all progress. Even a thought far removed from the common highway of thinking may have value for a future development and should never be discarded simply because it is "strange".

Second, that recent developments in public affairs—too hot and burning in our minds to be called historical facts as yet—teach us in a thundering voice that our past way of thinking (based, for the most part, on the methods of a mechanical age) has not been adequate to reality. We have talked about peace without being able to maintain it. Human coordination has not yet reached the point of forming a humanity which is a living unit with functioning members and organs. Instead, the organs of this body, humanity, fight one against the other. There has been a great deal of talk about economic systems and their relationships, both inside and outside the nation, but it has been either the economic crisis or the insufficient production of urgently needed commodities which has dominated the discussions. During the last few decades the thoughts accompanying and supporting a great desire for a free world-trade were too weak to hinder a complete breakdown of world-economics.

Now it is not the intention of this article to complain or criticise or count out mistakes, but only to suggest a new point of view, a new spiritual orientation. For this is the most pressing need of our time: to

126

develop thoughts adequate to ever changing reality, thoughts bearing within themselves upbuilding, future-developing forces.

Let us take one factor in economic life, money, and apply the functional, biological, dynamic method to the problem it presents. Question: Should human beings work for money or work for the work itself? It is a fundamental question—probably **the** question which will finally decide whether or not there will ever be a solution to our economic troubles.

That human labour can be paid for, is for sale, was a disagreeable idea to the Marxian system with its purely materialistic ideology, which has extended its shadow over this century too.

It is true that, for many human beings, money is a major incentive to working and striving. It is also true that in a human sense that is an unworthy state of affairs. Human culture, our highest good and ideal, consists of the sum total of the progress made in science, art and religion, that is, in all fields of human activity. Every thought devoted **only** to money-making is deducted from man's cultural development. Furthermore, the human being as an individual participates in eternity, having derived from a spiritual world and, as far as the ego or deepest self is concerned, surviving after death. We cannot take money with us. Therefore all this striving for money is a largely vain part of our lives which, with the last disease or suffocation, will lose meaning. In other words, its value to the individual is limited in time. The materialist, however, does not believe in an eternal life for the human ego, often considers money **the** thing, since after his death it can be transmitted to his offspring, thus physically extending, so to speak, the results of his existence. But biographical history shows that this is a rather short extension since in few families does the possession of a large amount of capital last over four generations, as contrasted to objects of art, the findings of science and the fundamental thoughts of philosophy, which endure for ages. Thus only that money which, indirectly, through people's contributions, support the creative part of human nature, helps to further cultural progress. Money fulfills its task only when it is consumed. Stored money, capital, is inactive, dead, and as a consequence will either be lost or become destructive. Only as long as it is productive is it healthy.

Comparing the economic body with a real living being and its laws with the laws of life, we find only two functions which involve something constantly produced in order to be consumed: the functions supported by food and red blood. Food comes from the outside, yet stimulates and supports the body's life just as productive ideas nourish the economic process. Blood, on the other hand, is entirely from within, circulating, being constantly produced and consumed in order to maintain life, supporting the organic functions necessary to growth and reproduction. It is important to realize that the blood is produced by the same organism by which it is consumed; that it is rarely stored and then only in small amounts, for instance, in the spleen; that it is in constant circulation and (this is probably its greatest secret) is **properly** (not evenly) distributed wherever it is needed. This is the real concept of a biologically sound and healthy unit of organization.

Money, curiously, fulfills a similar function; that is, it is or should be in constant circulation; it supports and maintains economic life and growth; it is produced by the same body economic by which it is consumed; and should—and here we touch upon the weakness and disease of our present economic system— not be stored for long or in large quantities (we could speak of an economic haematoma or, in a boom, of its opposite, an economic haemorrhage); and should be properly (not evenly) distributed wherever it is really needed. There should be no anaemia of a vital organ, but also no excess.

Finally, the proper distribution of blood is what gives us an understanding of an organism or biological unit and its functions, because there is a higher constructive idea behind it, a higher wisdom of life forces—higher than our intellectual, analytical thought can grasp.

The proper distribution of money—which alone can maintain a healthy economic body—has to be governed by a body of wisdom, that is, the concentrated economic and cultural wisdom of a nation directed by a mutually acceptable idea. In a free country, this body would have an advisory function, but it would be against common sense and self-destructive not to follow the advice.

Here a new thought enters our consideration. We neither possess nor are possessed by the red blood. (Only an unreal and defective philosophy such as the "blood and soil" idea of Germany could claim that man is possessed by the blood.) We merely **use blood; it is a** means of sustenance, we do not "own" it, but our system has to circulate it, to permit a constant new creation and consumption. The analogy reveals an important, basic function of money. There exists a vital obligation to do all in our power to keep it fluid and active. It is the dynamic process.

What would happen to our economic system if we accepted this point of view? We would no longer sell our work in order to sustain life. Nor would we be possessed by our money. And the sting of the thought, "working for money", would be removed because we would be working for a higher idea: the principal basic constructive idea that human beings are involved in a system of mutual collaboration, each working for the other, exchanging the products of their labour for the satisfaction of their mutual needs. To work is a human as well as economic necessity insofar as we participate in it. We do not work for money, but we do work: first, because it is to our own interest to do so in order to support the economic body; second, because in doing so we experience the fact that we are parts, members, organs of this body economic. If the heart goes wrong, the brain and lungs and everything suffer. And so it is when the single individual goes wrong in the sense of not finding his place and task (function) in the process of economic production and consumption. All the other members are affected.

I am working not because I need a living or make so much money that I can store it and transform it into capital, but I am working because I feel I am sharing in the general economic welfare because it is a satisfaction to know that I am a valuable, an indispensable member of the social organism, because it is a satisfaction to be "functioning". It is up to the wisdom of social leadership, manifesting first in education and involving a fundamental philosophy of life, to teach me to reorganise my abilities and possibilities and help me find my proper place. The useless in such an organically balanced system would soon become deprived of the unnecessarily stored "red cells" above his needs. On the other hand the one without possessions but economically productive would always find support for his needs because he is part of the economic organism and it is to the interest of the totality that he neither suffer nor perish. His maintenance is just as important to the entire organism as the maintenance of the organism is important to him.

Fundamentally, this is the application of the original Christian thought, a moral power representing humanity as such. From now on the Christian morality must govern not only our thoughts and feelings and religious desires but everything we do as the result of an activity of will in collaboration with and out of an understanding of other human beings, in mutual free agreement.

To think this does not harm, even if the thought sounds "strange" and "idealistic". History teaches us that, throughout the centuries to find this organic solution, man has made many approaches and failures. And there will be many more. As long as we do not find peace within we cannot expect peace without. As long as we do not realize that we work because we love our work and we love to work, inasmuch as working is the highest expression of human dignity, so long as we do not realize this, we will not be able to find other thoughts tending toward the improvement of the economic system.

Then this thought must become part of our heart, of that never ceasing motor of energy which gives us the power to change conditions, to transform the world. The first and only thing we have to do is change our attitude toward ourselves. I am I, not because of me, because I am the closest and nearest relative of myself, but because I am the bearer of an eternal entity—just as good and just as bad as any other human being, with equal rights and equal responsibilities and obligations. I have to maintain and improve myself not only in my interest but, even more so, in the interest of the other, the community of which the economic process is a part. As for the other being, he is working for me just as much as I for him. We have to give up our egocentric point of view. This, however, may be the greatest and most difficult task ever given to humanity. But once we have taken this step, we shall discover that it yields the greatest possible satisfaction, not only compensating us for all our efforts and pains, but conferring upon us an ever increasing moral power and truth. This is what is most needed in economic life; a moral point of view. Because the human being is entrusted with the power of life and death, of growth and destruction, the future history of mankind will present the balance of this responsibility.

Once these thoughts have become power it will not be difficult to transform them into deeds, to find adequate administrative measures to build up a new and healthier body social. The fact of their becoming power would remove at once the spiritual cause of this dreadful war. May it be a point in a program for a future peace: work is an expression of human dignity, and satisfies the human part of the human being.

It is man's participation in a healthy social life which raises him above the animal kingdom. Animals act only in behalf of themselves: take care of their offspring, search for food and shelter, and—at least this is true of many of them—engage in mutual destruction. Animals have no economic system and no money. This is up to the human race, but not in order to continue practices common and unavoidable in the animal realm. For the purpose of creation is this: to create once again, that is, to create values which are of longer duration and further reaching than a single lifetime, and involve more than one individual.